Generalized Cauchy–Riemann systems
with a singular point

π Pitman Monographs and
Surveys in Pure and Applied Mathematics 85

Generalized Cauchy–Riemann systems with a singular point

Z D Usmanov

Institute of Mathematics,
Tajik Academy of Sciences

LONGMAN

Addison Wesley Longman Limited
Edinburgh Gate, Harlow
Essex CM20 2JE, England
and Associated companies throughout the world.

*Published in the United States of America
by Addison Wesley Longman Inc.*

First published 1997

AMS Subject Classifications: (Main) 35J, 35C, 45P
(Subsidiary) 30E, 45E, 53C

ISSN 0269-3666

ISBN 0 582 29280 8

British Library Cataloguing in Publication Data

A catalogue record for this book is
available from the British Library

Library of Congress Cataloging-in-Publication Data

Usmanov, Zafar Dzuraevich.
Generalized Cauchy–Riemann systems with a singular point / Zafar Dzuraevich Usmanov.
p. cm. -- (Pitman monographs and surveys in pure and applied mathematics, ISSN 0269-3666 ; ???)
ISBN 0-582-29280-8 (alk. paper)
1. CR submanifolds. 2. Singularities (Mathematics) I. Title.
II. Series.
QA649.U86 1997
516.3'62--DC20 96-31177
CIP

Printed and bound by Bookcraft (Bath) Ltd

Contents

Preface

In this monograph a theory of generalized Cauchy–Riemann systems with polar singularities of order not less than 1 is presented and its application to the study of infinitesimal bendings of positive curvature surfaces with an isolated flat point is given here. It contains results of investigations obtained recently by the author and his collaborators. In this monograph special attention is paid to the description of formal methods of constructing general integral operators which are a natural extension of the classical apparatus of generalized analytic function theory.

The monograph is written not only for specialists in complex analysis, geometry and mechanics but also for student–mathematicians who could use it as a manual.

Professor Zafar D. Usmanov
Institute of Mathematics
of the Tajik Academy of Sciences
ul. Aini 299, Academgorodok
Dushanbe 734 063
TAJIKISTAN

Introduction

The elliptic system of first–order partial differential equations of the following type

$$\begin{aligned} \frac{\partial \mathrm{u}}{\partial x} - \frac{\partial v}{\partial y} + a\mathrm{u} + bv = 0, \\ \frac{\partial \mathrm{u}}{\partial y} - \frac{\partial v}{\partial x} + c\mathrm{u} + dv = 0, \end{aligned} \tag{1}$$

is called a generalized Cauchy–Riemann system. Here u and v are unknown and a, b, c, d are given functions of the real variables x and y in some domain G. This system extends the classical Cauchy–Riemann system

$$\frac{\partial \mathrm{u}}{\partial x} - \frac{\partial v}{\partial y} = 0, \quad \frac{\partial \mathrm{u}}{\partial y} - \frac{\partial v}{\partial x} = 0, \tag{2}$$

which can be obtained from (1) for $a = b = c = d = 0$.

If we introduce the complex variable $z = x + \mathrm{i}y$, $\mathrm{i}^2 = -1$, the complex unknown function $w = \mathrm{u} + \mathrm{i}v$, and the notation

$$4A(z) = a + d + \mathrm{i}c - \mathrm{i}b,$$

$$4B(z) = a - d + \mathrm{i}c + \mathrm{i}b,$$

$$\partial_{\bar{z}} = \frac{1}{2}\left(\frac{\partial}{\partial x} - \mathrm{i}\frac{\partial}{\partial y}\right),$$

then (1) and (2) can be written as

$$\begin{aligned} \partial_{\bar{z}} w + A(z)w + B(z)\bar{w} = 0, \\ \partial_{\bar{z}} w = 0. \end{aligned} \tag{3}$$

The theory of $w(z)$–functions, $z \in G$, satisfying (3), was constructed by L. Bers and I.N. Vekua assuming that $A(z)$ and $B(z)$ belong to $L_p(G), p > 2$, [69].[1] In this

[1] N. Bliev has widened the theory of generalized analytic functions on fractional spaces [13].

case (3) is called the *regular generalized Cauchy–Riemann system* and its solutions are named *generalized analytic functions.* Such system coefficients can allow small singularities. In particular, if $A(z)$ and $B(z)$ tend to infinity at some isolated point of G, then the order of the singularity at this point must be strictly less than 1. I. Vekua has introduced generalized Cauchy–Riemann systems with *quasi-summarized coefficients* $A(z)$ and $B(z)$. This means that there exist analytic functions $\varphi(z)$ and $\psi(z), z \in G$, of the complex variable z such that φA and $\psi B \in L_p(G), p > 2$. For the solutions of this type of system Vekua has proposed the formula

$$w(z) = \Phi(z)e^{\omega(z)}, \tag{4}$$

where $\Phi(z)$ is an analytic function in G and

$$\omega(z) = \frac{1}{\pi\varphi(z)} \iint_G \frac{\varphi(\zeta)A(\zeta)}{\zeta - z}\, d\xi\, d\eta + \frac{1}{\pi\psi(z)} \iint_G \frac{\psi(\zeta)B(\zeta)}{\zeta - z} \frac{\overline{w(\zeta)}}{w(\zeta)}\, d\xi\, d\eta, \tag{$*$}$$

This formula establishes a connection between the sets of solutions of the system (3) with quasi–summarized coefficients and analytic functions of a complex variable. Namely, if $w(z)$ is a given solution of the system (3), then the analytic functions $\Phi(z)$ which corresponds to $w(z)$ is uniquely determined by (4). And conversely, for a given analytic function $\Phi(z)$ the concrete $w(z)$ can be determined [70].

The formula (4) becomes especially effective for $B(z) \equiv 0$, when $w(z)$ does not depend on $w(z)$. In this case it gives the general solution of (3) and can be applied specifically to the investigation of the structure of the solution in the neighbourhood of the singular point of $A(z)$ [71].

For $B(z) \neq 0$ the application of the formulas (4), ($*$) encounters serious obstacles. The fact is that the prior information on the asymptotic behaviour $w(z)$ at singular points of the coefficients $A(z)$ and $B(z)$ is unknown. Therefore it is impossible to estimate the behaviour of the second term in ($*$) and $\omega(z)$ in (4). Apparently this situation is a common reason for not using the formula (4) as a tool for the investigation of singular generalized Cauchy–Riemann systems. However, if we take into consideration that in ($*$), besides the second term there is the first one as well, and we attempt to dispose of the singularities in the coefficients of equation (3) so that the basic role in the asymptotic behaviour of $\omega(z)$ could be played by the first term, then the representative (4), ($*$) can be efficient.

It is this very approach that contributed to the progress that was achieved by G. Makatsaria [32, 33]. In his paper

$$r^{\nu} A(z) = a(z)e^{im\varphi}, \qquad r^{\mu} B(z) = b(z),$$

where $z = re^{i\varphi}, z = 0 \in G, m$ is integer, $\nu > 1$ and $\mu \geq 0$ are real numbers; $(\alpha(z) - \alpha(0))r^{-\alpha} \in L_p(G)$ and $b(z) \in L_q(G)$; moreover $p, q > 2$ and $\alpha > \nu - [\nu]$.

If we set $\varphi(z) = z^{[\nu]}, \psi(z) = z^{[\mu]}$ and consider the additional restrictions $m \neq 1, m \neq \nu, m + \nu \neq 2, [\nu] \geq \mu + 1$, then the first term in formula (*) will have a singularity at $z = 0$ higher than the second one. For a circular neighbourhood of the point $z = 0$ it is not difficult to calculate the value of the first term and consequently to define the asymptotic expression of $\omega(z)$. Now a *priori* the structure of $\omega(z)$– functions in a neighbourhood of the singular point $z = 0$ has become known. This allows us to obtain some results on properties of the equation (3) with quasi–summarized coefficients, in particular to prove the absence of non–trivial solutions in the class of functions, admitting at $z = 0$ a pole of an arbitrarily high order.

For $B(z) \not\equiv 0$ the information about quasi–summability of $A(z)$ and $B(z)$ is too common and we cannot use it for the development of the theory of (3). For this reason other authors have concentrated their research on special equations. First of all it concerns the equation

$$\partial_{\bar{z}} w + \frac{a(z)}{|z|} w + \frac{b(z)}{|z|} \bar{w} = 0, \quad z \in G, \tag{5}$$

considered by L.G. Mikhailov [36]. This equation is the unique representative of the generalized Cauchy–Riemann systems with quasi–summarized coefficients in which the fundamental operator of the regular generalized Cauchy–Riemann systems theory

$$T(\cdot) = -\frac{1}{\pi} \iint_G \frac{(\cdot)}{\zeta - z}\, d\xi\, d\eta, \quad \zeta = \xi + i\eta, \tag{6}$$

can be applied. Assuming $a(z)$ and $b(z)$ are bounded measurable functions in G and using (6) in (5), L.G. Mikhailov derives the integral equation

$$w(z) - \frac{1}{\pi} \iint_G \frac{a(\zeta)w(\zeta) + b(\zeta)\overline{w(\zeta)}}{|\zeta|(\zeta - z)}\, d\xi\, d\eta = \Phi(z), \tag{7}$$

where $\Phi(z), z \in G$, is an analytic function of z. The operator in (7) is bounded in the special classes of functions admitting a small polar singularity at $z = 0$ [35, 36].

The solution of (7) exists and is unique provided that the operator has a small norm. This is equivalent to the smallness of $|\alpha(z)|$ and $|b(z)|$. In this case there exists a one–to–one correspondence between the elements of the sets $\{w(z)\}$ and $\{\Phi(z)\}$.

Other authors, using the assumption that $\alpha(z)$ and $b(z)$ are analytic functions of z and $\bar{z}$, studied the existence of analytic solutions with respect to z and $\bar{z}$ in a neighbourhood of the point $z = 0$.

This monograph offers a unified approach to constructing solutions of the equation

$$\partial_{\bar{z}} w + \frac{e^{i(n+1)\varphi}}{2} \frac{b(z)}{r^{1+\nu}} \bar{w} = F(z), \quad z \in G, \tag{8}$$

where $z = x + iy = re^{i\varphi}, n$ is some integer, ν is a real number and $b(z)$ and $F(z)$ are given functions; moreover $b(z)$ is continuous in G and, besides the $b(0) \neq 0$ condition is subordinated to an additional requirement in a neighbourhood of $z = 0$. This equation does not contain the term $w(z)$, but this is not of much importance. In fact, by substituting $w_1(z) = \gamma(z)w(z)$, where $\gamma(z)$ is a solution of the equation $\partial_{\bar{z}}\gamma = A(z)\gamma$, the new unknown function $w_1(z)$ will satisfy the equation of type (3) in which the coefficient A is equal to zero.

To explain the main idea for the investigation of equation (8) it is necessary to say the following. It is known that the development of the theory of generalized analytic functions is closely connected with the application of the operator (6). On the basis of (6) integral representations of the first and second types are obtained which establish a correspondence between generalized analytic functions and analytic functions of a complex variable [69]. As was noted earlier, the operator is also used to study the singular system (5). However, if in the regular case this operator possesses rather 'good' properties, then even for the simplest singular system (5), the above operator gives the integral equation (7) with an unbounded operator in the class of continuous functions. This fact is easily explained: the application of the operator (6) is assumed to have the description of the properties of the singular system solutions using analytic functions of z, and thus interpreting the analytic functions as a convenient model of the solutions of the singular systems. But it is necessary to conclude that this simulation is not suitable because it is realized by an operator unbounded in the class of continuous functions.

In this connection the study of generalized Cauchy–Riemann systems with quasi–summarized coefficients must be based on more 'suitable' integral operators taking into account the characteristics of the equations. It is clear in this case that the theory of singular generalized Cauchy–Riemann systems will not be associated with analytic functions, but with a theory of specific model equations with singular coefficients. The essence of the unified approach in this monograph is in the development of a theory of such equations similar to the theory of analytic functions of a complex variable, in constructing 'convenient' integral operators, and with their help in reducing properties of model solutions equations to general ones.

In chapters *1–3* this approach is thoroughly demonstrated on equation (8) for $n = \nu = 0$. Chapter 4 considers the case $n \neq 0$ and $\nu = 0$. Here the corresponding results are given mainly in concise form. Chapter 5 is devoted to the study of equation (8) with a singularity of more than order 1 ($\nu \neq 0$). The analysis of every case separately depends on the distinction of the properties of the solution and the analytic operators used. For example, the kernels of the basic integral operator for $n = \nu = 0$ are given by infinite series composed of power function. The analogous representation for $n \neq 0$ and $\nu \neq 0$ is more bulky, and for $\nu \neq 0$ is more complex since the series terms are products of Bessel functions. It should be noted that in the 'Supplement' at the end of the book, where the generalized Cauchy–Riemann system with a singular line is discussed, the appropriate kernels are expressed by series of hypergeometric functions. Extracting of the main singularity in kernels is connected with certain technical obstacles and requires the utilization of specific analytic methods in every case.

It is also necessary to emphasize that the analytic apparatus elaborated here is a natural extension of generalized analytic function methods to the study of singular generalized Cauchy–Riemann systems. It is sufficient to mention that the basic integral operators constructed for singular systems, applied to the regular case, are transformed into the operator (8).

The successive theories of singular and regular systems in Chapter 6 is discussed as well. It is established here that the equations of Chapters *1–4* arise naturally. They are intended for the investigation of infinitesimal bendings of positive curvature surfaces with a specific structure in a neighbourhood of an isolated flat point. For more generalized structures a singular equation has been obtained, but we have not yet succeeded in studying it. In this connection in many chapters we formulate unsolved problems and hope that they will attract the attention of researchers.

Chapter 1
Interrelation between sets of general and model equation solutions

The basic aim of this chapter is to establish a connection between the solutions of the equations

$$\partial_{\bar{z}} w - \frac{b(z)}{2\bar{z}} \bar{w} = F(z), \quad z \in G,$$
$$\partial_{\bar{z}} \Phi - \frac{b(0)}{2\bar{z}} \bar{\Phi} = 0, \quad z \in G,$$

by means of a linear integral equation with a completely continuous operator. Thanks to this the study of a certain question for the general equation is reduced to an analogous question for the equation of the particular type.

The results of this chapter have been obtained in [52, 54, 63].

1 The method of constructing a general integral operator

We shall consider the equation

$$\partial_{\bar{z}} w - \frac{\lambda}{2\bar{z}} \bar{w} = f(z), \quad z \in G, \tag{1.1}$$

where $w(z)$ is an unknown and $f(z)$ is a given function, $\lambda = \lambda_1 + i\lambda_2$ is a complex number, G is a simply connected domain and $z = 0 \in G$.

The immediate purpose of this section is a formal construction of a general solution for (1.1). It is clear that this solution must have the following form

$$w(z) = \Phi(z) + S_G f, \quad z \in G, \tag{1.2}$$

i.e. it contains the particular solution $S_G f$ of (1.1) and the general solution $\Phi(z)$ of the homogeneous equation

$$\partial_{\bar{z}} \Phi - \frac{\lambda}{2\bar{z}} \bar{\Phi} = 0, \quad z \in G. \tag{1.3}$$

Let G be the disk $|z| < R$. Assuming $z = re^{i\varphi}$ and rewriting (1.1) in polar coordinates, we get

$$\frac{\partial w}{\partial r} + \frac{i}{r}\frac{\partial w}{\partial \varphi} - \frac{\lambda}{r}\bar{w} = 2e^{-i\varphi} f(z). \tag{1.4}$$

Without mentioning the classes of $w(z)$ and $f(z)$ we multiply two parts of this equation by $e^{-ik\varphi}/2\pi, k = 0, \pm 1 \ldots$, and integrate with respect to φ from 0 to 2π. If we introduce the notation

$$w_k(r) = \frac{1}{2\pi}\int_0^{2\pi} w(z)e^{-ik\varphi}\, d\varphi, \qquad f_k(r) = \frac{1}{2\pi}\int_0^{2\pi} f(z)e^{-i(k+1)\varphi}\, d\varphi, \tag{1.5}$$

then the above operations lead to an ordinary differential equation for defining an unknown function $w_k(r)$ which is the kth Fourier coefficient of the unknown solution $w(z)$:

$$\frac{dw_k}{dr} - \frac{k}{r}w_k - \frac{\lambda}{r}\bar{w}_-k = f_k(r). \tag{1.6}$$

For $k = 0$ we have

$$\frac{dw_0}{dr} - \frac{\lambda}{r}\bar{w}_0 = f_0(r)$$

The result of integrating this equation depends on λ. For $\lambda > 0$

$$w_0(r) = -\int_r^R \left(\frac{r}{\rho}\right)^\lambda \operatorname{Re} f_0\, d\rho + i\int_0^r \left(\frac{\rho}{r}\right)^\lambda \operatorname{Im} f_0\, d\rho + a_0 r^\lambda + ib_0 r^{-\lambda}. \tag{1.7}$$

For $\lambda < 0$

$$w_0(r) = \int_0^r \left(\frac{r}{\rho}\right)^\lambda \operatorname{Re} f_0\, d\rho - i\int_0^r \left(\frac{\rho}{r}\right)^\lambda \operatorname{Im} f_0\, d\rho + a_0 r^\lambda + ib_0 r^{-\lambda}. \tag{1.7$_-$}$$

For $\lambda_2 = \operatorname{Im}\lambda \neq 0$

$$\begin{aligned} w_0(r) = &-\frac{i(|\lambda| - \lambda)}{2\lambda_2|\lambda|}\int_r^R \left(\frac{r}{\rho}\right)^{|\lambda|} \operatorname{Re}\left[(|\lambda| + \bar{\lambda})f_0\right] d\rho \\ &-\frac{i(|\lambda| + \lambda)}{2\lambda_2|\lambda|}\int_0^r \left(\frac{\rho}{r}\right)^{|\lambda|} \operatorname{Re}\left[(|\lambda| - \bar{\lambda})f_0\right] d\rho \\ &+i(|\lambda| - \lambda)a_0 r^{|\lambda|} + i(|\lambda| + \lambda)b_0 r^{|\lambda|}. \end{aligned} \tag{1.7$_i$}$$

In these formulas a_0 and b_0 are arbitrary real constants.

Let's consider (1.6) for $k \neq 0$. This equation contains two unknown functions w_k and w_{-k}. If we substitute$-k$ for k, then we obtain

$$\frac{dw_{-k}}{dr} + \frac{k}{r}w_{-k} - \frac{\lambda}{r}\bar{w}_k = f_{-k}(r), \tag{1.8}$$

which with (1.6) comprises the system of equations for determining w_k and w_{-k}. Integrating (1.6) and (1.8), we get

$$\begin{aligned} w_k(r) = &-\frac{1}{2\mu_k}\int_r^R \left(\frac{r}{\rho}\right)^{\mu_k}(P_k f_k + \lambda \bar{f}_{-k})\,\mathrm{d}\rho \\ &+\frac{1}{2\mu_k}\int_0^r \left(\frac{\rho}{r}\right)^{\mu_k}(P_{-k} f_k - \lambda \bar{f}_{-k})\,\mathrm{d}\rho + \lambda a_k r^{\mu_k} + \lambda b_k r^{-\mu_k}, \end{aligned} \tag{1.9}$$

$$\begin{aligned} w_{-k}(r) = &-\frac{1}{2\mu_k}\int_r^R \left(\frac{r}{\rho}\right)^{\mu_k}(\lambda \bar{f}_k + P_{-k} f_{-k})\,\mathrm{d}\rho \\ &-\frac{1}{2\mu_k}\int_0^r \left(\frac{\rho}{r}\right)^{\mu_k}(\lambda f_k - P_k f_{-k})\,\mathrm{d}\rho + P_{-k}\bar{a}_k r^{\mu_k} - P_k \bar{b}_k r^{-\mu_k}, \end{aligned}$$

In these formulas a_k and b_k are arbitrary complex constants and

$$\mu_k^2 = |\lambda|^2 + k^2, \qquad P_k = \mu_k + k, \qquad P_{-k} = \mu_k - k, \qquad k = 1, 2, \ldots . \tag{1.10}$$

Adapting (1.5) in formulas (1.7), (1.9), the functions $f_0(\rho), f_k(\rho), f_{-k}(\rho)$ must be expressed through $f(\zeta), \zeta = \rho e^{i\gamma}$ (we do not extract these formulas on account of their complexity). Then the expressions obtained are substituted for $w_0(r), w_k(r)$ and $w_{-k}(r)$ in the Fourier series

$$w(z) = \sum_{k=-\infty}^{\infty} w(r)e^{ik\varphi}.$$

Finally the orders of integration and summation are formally changed. This procedure results in (1.2), in which

$$\begin{aligned} \Phi(z) = &A_0 r^{|\lambda|} + B_0 r^{-|\lambda|} \\ &+\sum_{k=1}^{\infty}\left[\lambda(a_k r^{\mu_k} + b_k r^{-\mu_k})e^{ik\varphi} + (P_{-k}\bar{a}_k r^{\mu_k} - P_k \bar{b}_k r^{-\mu_k})e^{-ik\varphi}\right], \end{aligned}$$

the constants A_0 and B_0 in the first two terms being subject to λ and being expressed in different ways through a_0 and b_0; see $(1.7_+), (1.7_-), (1.7_i)$:

$$S_G f = -\frac{1}{\pi}\iint_G \left[\frac{\Omega_1(z,\zeta)}{\zeta} f(\zeta) + \frac{\Omega_2(z,\zeta)}{\bar{\zeta}}\overline{f(\zeta)}\right] \mathrm{d}\xi\,\mathrm{d}\eta, \tag{1.11}$$

where $\zeta = \xi + \mathrm{i}\eta = \rho\mathrm{e}^{\mathrm{i}\gamma}$, and Ω_1 and Ω_2 are defined by the formulas:

$$\Omega_1 = \begin{cases} \dfrac{1}{2}\left(\dfrac{r}{\rho}\right)^{|\lambda|} + \displaystyle\sum_{k=1}^{\infty} \dfrac{P_k \mathrm{e}^{\mathrm{i}k(\varphi-\gamma)} + P_{-k}\mathrm{e}^{-\mathrm{i}k(\varphi-\gamma)}}{2\mu_k}\left(\dfrac{r}{\rho}\right)^{\mu_k}, & |z| < |\zeta|, \\ -\dfrac{1}{2}\left(\dfrac{\rho}{r}\right)^{|\lambda|} - \displaystyle\sum_{k=1}^{\infty} \dfrac{P_{-k} \mathrm{e}^{\mathrm{i}k(\varphi-\gamma)} + P_{k}\mathrm{e}^{-\mathrm{i}k(\varphi-\gamma)}}{2\mu_k}\left(\dfrac{\rho}{r}\right)^{\mu_k}, & |z| > |\zeta|, \end{cases}$$

$$(1.12)$$

$$\Omega_2 = \begin{cases} \dfrac{\lambda}{2|\lambda|}\left(\dfrac{r}{\rho}\right)^{|\lambda|} + \displaystyle\sum_{k=1}^{\infty} \dfrac{\lambda}{\mu_k}\left(\dfrac{r}{\rho}\right)^{\mu_k} \cos k(\varphi-\gamma), & |z| < |\zeta|, \\ \dfrac{\lambda}{2|\lambda|}\left(\dfrac{\rho}{r}\right)^{|\lambda|} + \displaystyle\sum_{k=1}^{\infty} \dfrac{\lambda}{\mu_k}\left(\dfrac{\rho}{r}\right)^{\mu_k} \cos k(\varphi-\gamma), & |z| > |\zeta|. \end{cases}$$

Thus a formal way we have derived relation (1.2) which, however, we should not consider finally proved. But we have achieved our basic aim of having written out the fundamental operator S_G in explicit form.

2 Properties of the functions Ω_1 and Ω_2

Beginning with this section, we shall verify relation (1.2). First of all the problem is about (1.12). The functions Ω_1 and Ω_2 of two complex variables z and ζ on the direct product $E \times E$ of two complex planes E are worth considering.

2.1 If we have a careful look at (1.12) we shall see that the points $z = \zeta = 0$ and $z = \zeta = \infty$ should be excluded from the domain of definition of the function as kernel values in them are not defined. Besides it is necessary to analyse in detail the behaviour of $\Omega_1(z,\zeta)$ and $\Omega_2(z,\zeta)$ at $|z| = |\zeta|$, as on this set, in every line of the right side of (1.12) there are series that may diverge. For instance, in the upper line ($|z| < |\zeta|$) for Ω_1 the divergent series is

$$\sum_{k=1}^{\infty} \frac{P_k}{2\mu_k}\mathrm{e}^{\mathrm{i}k(\varphi-\gamma)}\left(\frac{r}{\rho}\right)^{\mu_k},$$

where, by virtue of adapted notation, we have $\mu_k^2 = |\lambda|^2 + k^2$ and $P_k = \mu_k + k$.

According to the above, further study of the properties of Ω_1 and Ω_2 will be simultaneously investigated by emphasizing their main singularities on the set $|z| = |\zeta|$, more precisely for $z = \zeta$.

Lemma 2.1 *For $(z,\zeta) \in E \times E, \Omega_1(z,\zeta)$ is represented in the form*

$$\Omega_1(z,\zeta) = \frac{\zeta}{\zeta - z} + \Omega_1^0(z,\zeta), \tag{2.1}$$

where Ω_1^0 is continuous in the two variables z and ζ everywhere except the points $z = \zeta = 0$ and $z = \zeta = \infty$. *For a fixed ζ (or z) different from zero and infinity Ω_1* for $z = 0$ and $z = \infty$ *(respectively for $\zeta = 0$ and $\zeta = \infty$) has a zero of order $|\lambda|$.*

First we shall prove the last statement of the lemma. Let us fix for example $\zeta = \zeta_0 = \rho_0 e^{i\gamma_0}$ and rewrite Ω_1 in the form

$$\Omega_1 = \begin{cases} \left(\dfrac{r}{\rho_0}\right)^{|\lambda|} \Omega_1'(z,\zeta_0), & |z| < |\zeta_0| \\ \left(\dfrac{\rho_0}{r}\right)^{|\lambda|} \Omega_1''(z,\zeta_0), & |z| > |\zeta_0| \end{cases} \tag{2.2}$$

where according to (1.12)

$$\Omega_1'(z,\zeta_0) = \frac{1}{2} + \sum_{k=1}^{\infty} \frac{P_k e^{ik(\varphi-\gamma_0)} + P_{-k} e^{-ik(\varphi-\gamma_0)}}{2\mu_k} \left(\frac{r}{\rho_0}\right)^{\mu_k - |\lambda|},$$

$$\Omega_1''(z,\zeta_0) = -\frac{1}{2} - \sum_{k=1}^{\infty} \frac{P_{-k} e^{ik(\varphi-\gamma_0)} + P_k e^{-ik(\varphi-\gamma_0)}}{2\mu_k} \left(\frac{\rho_0}{r}\right)^{\mu_k - |\lambda|}.$$

Taking the notation (1.10) into account, it is easy to prove that Ω_1' and Ω_1'' are continuous with respect to z for $z = 0$ and $z = \infty$ respectively. Then from (2.2) it follows that Ω_1 has a zero of order $|\lambda|$ for $z = 0$ and $z = \infty$.

In the same way, Lemma 2.1 about the zeros of Ω_1 with respect to ζ for fixed $z = z_0$ is proved.

Now we prove the coninuity of Ω_1^0 separately in the domains $|z| \geq |\zeta|$ and $|z| \leq |\zeta|$ and the absence of a discontinuity on the set $|z| = |\zeta|$. Let us use the identity

$$\frac{\zeta}{\zeta - z} \equiv \begin{cases} \displaystyle\sum_{k=0}^{\infty} \left(\frac{r}{\rho}\right)^k e^{ik(\varphi-\gamma)}, & |x| < |\zeta|, \\ \displaystyle -\sum_{k=1}^{\infty} \left(\frac{\rho}{r}\right)^k e^{-ik(\varphi-\gamma)}, & |z| > |\zeta|. \end{cases}$$

By means of this Ω_1^0 for $|z| < |\zeta|$ can be written in the form

$$\begin{aligned}\Omega_1^0 =& \frac{1}{2}\left(\frac{r}{\rho}\right)^{|\lambda|} - 1 + \sum_{k=1}^{\infty}\left[\frac{P_k}{2\mu_k}\left(\frac{r}{\rho}\right)^{\mu_k} - \left(\frac{r}{\rho}\right)^{k}\right]e^{ik(\varphi-\gamma)} \\ &+ \sum_{k=1}^{\infty}\frac{P_{-k}}{2\mu_k}\left(\frac{r}{\rho}\right)^{\mu_k} e^{-ik(\varphi-\gamma)}.\end{aligned} \tag{2.3}$$

The two sums in this relation are uniformly convergent series for $|z| \le |\zeta|$. This follows from the Weierstrass test, using the inequality

$$\begin{aligned}&\left|\left(\frac{r}{\rho}\right)^{k} - \frac{P_k}{2\mu_k}\left(\frac{r}{\rho}\right)^{\mu_k}\right| < \left(\frac{2k}{P_k}\right)^{k/P_{-k}}\frac{\mu_k - k}{\mu_k} < \frac{|\lambda|^2}{4k^2}, \\ &\left|\frac{P_{-k}}{2\mu_k}\left(\frac{r}{\rho}\right)^{\mu_k}\right| < \frac{|\lambda|^2}{4k^2},\end{aligned} \tag{2.4}$$

Here we have applied notation (1.10) and the relations $P_k \cdot P_{-k} = |\lambda|^2, \mu_k > k$.

Now the continuity of Ω_1^0 in the domain $|z| \le |\zeta|$ is a trivial consequence of the uniform convergence of the above series all terms of which for $\zeta \ne 0$ are continuous in the two variables z, ζ in the domain $|z| \le |\zeta|$.

In a similar way the continuity of Ω_1^0 is verified for $|z| \ge |\zeta|$, the basis being the equality

$$\begin{aligned}\Omega_1^0 =& -\frac{1}{2}\left(\frac{\rho}{r}\right)^{|\lambda|} - \sum_{k=1}^{\infty}\frac{P_{-k}}{2\mu_k}\left(\frac{\rho}{r}\right)^{\mu_k} e^{ik(\varphi-\gamma)} \\ &- \sum_{k=1}^{\infty}\left[\frac{P_k}{2\mu_k}\left(\frac{\rho}{r}\right)^{\mu_k} - \left(\frac{\rho}{r}\right)^{k}\right]e^{-ik(\varphi-\gamma)}.\end{aligned} \tag{2.5}$$

It remains to check the continuity of Ω_1^0 on the set $|z| = |\zeta|$ which is the common boundary of the domains $|z| < |\zeta|$ and $|z| > |\zeta|$. But it follows from this fact that the values Ω_1^0 taken from formulas (2.3) and (2.5) for $|z| = |\zeta| (z = \rho)$ coincide.

Lemma 2.2 *The function $\Omega_2(z,\zeta)$ for $(z,\zeta) \in E \times E$* has the form

$$\Omega_2 = \Omega_2^0(z,\zeta) + \begin{cases}\lambda \cdot \ln\dfrac{|\zeta|}{|\zeta - z|}, & |z| \le |\zeta|, \\ \lambda \cdot \ln\dfrac{|z|}{|\zeta - z|}, & |z| \ge |\zeta|,\end{cases} \tag{2.6}$$

where Ω_2^0 is continuous in the two variables z and ζ everywhere except the points $z = \zeta = 0$ and $z = \zeta = \infty$.

For a fixed ζ (or z) not equal to zero and infinity, Ω_2 for $z = 0$ and $z = \infty$ (respectively for $\zeta = 0$ and $\zeta = \infty$) has zeros of order $|\lambda|$.

The last assertion is proved as in Lemma 2.1. Therefore we shall dwell upon the proof of the continuity of Ω_2^0.

In the domain $|z| < |\zeta|$ the function Ω_2^0 by the identity

$$\ln \frac{|\zeta|}{|\zeta - z|} \equiv \sum_{k=1}^{\infty} \frac{\cos k(\varphi - \gamma)}{k} \left(\frac{r}{\rho}\right)^k, \quad |z| < |\zeta|,$$

has the form

$$\Omega_2^0 = \frac{\lambda}{2|\lambda|}\left(\frac{r}{\rho}\right)^{|\lambda|} + \sum_{k=1}^{\infty} \lambda \left[\frac{1}{\mu_k}\left(\frac{r}{\rho}\right)^{\mu_k} - \frac{1}{k}\left(\frac{r}{\rho}\right)^k\right] \cos k(\varphi - \gamma). \tag{2.7}$$

All terms on the right are continuous in the variables z and ζ in the domain $|z| \le |\zeta|$. In the same domain the given series converges uniformly since

$$\left|\frac{1}{\mu_k}\left(\frac{r}{\rho}\right)^{\mu_k} - \frac{1}{k}\left(\frac{r}{\rho}\right)^k\right| < \frac{1}{k} - \frac{1}{\mu_k} < \frac{|\lambda|^2}{2k^3}.$$

Then Ω_2^0 is a continuous function for $|z| \le |\zeta|$.

In a similar way the continuity of Ω_2^0 in the domain $|z| \ge |\zeta|$ is established starting from the identity

$$\ln \frac{|z|}{|\zeta - z|} \equiv \sum_{k=1}^{\infty} \frac{\cos k(\varphi - \gamma)}{k} \left(\frac{\rho}{r}\right)^k, \quad |z| > |\zeta|,$$

and defining Ω_2^0 according to the equality

$$\Omega_2^0 = \frac{\lambda}{2|\lambda|}\left(\frac{\rho}{r}\right)^{|\lambda|} + \sum_{k=1}^{\infty} \lambda \left[\frac{1}{\mu_k}\left(\frac{\rho}{r}\right)^{\mu_k} - \frac{1}{k}\left(\frac{\rho}{r}\right)^k\right] \cos k(\varphi - \gamma). \tag{2.8}$$

By passing over from the domain $|z| < |\zeta|$ to the domain $|z| > |\zeta|$ the function Ω_2^0 does not undergo a discontinuity for the values of Ω_2^0 taken from (2.7) and (2.8), for $|z| = |\zeta|$ coincide.

2.2 Now we shall dwell upon the differential properties of Ω_1 and Ω_2. Define

$$\partial_{\bar z} = \frac{1}{2}\left(\frac{\partial}{\partial x} + \mathrm{i}\frac{\partial}{\partial y}\right), \qquad \partial_z = \frac{1}{2}\left(\frac{\partial}{\partial x} - \mathrm{i}\frac{\partial}{\partial y}\right).$$

In the polar coordinate system $r = |z|$ and $\varphi = \arg(z)$ we have

$$\partial_{\bar{z}} = \frac{e^{i\varphi}}{2}\left(\frac{\partial}{\partial r} + \frac{i}{r}\frac{\partial}{\partial\varphi}\right), \qquad \partial_z = \frac{e^{-i\varphi}}{2}\left(\frac{\partial}{\partial r} - \frac{i}{r}\frac{\partial}{\partial\varphi}\right).$$

On the basis of these formulas, the explicit forms of Ω_1 and Ω_2 and their properties noted in Lemmas 2.1 and 2.2, are established by direct verification:

$$\partial_{\bar{z}}\Omega_1 - \frac{\lambda}{2\bar{z}}\overline{\Omega_2} = 0, \qquad \partial_{\bar{z}}\Omega_2 - \frac{\lambda}{2\bar{z}}\overline{\Omega_1} = 0; \tag{2.9}$$

$$\partial_{\bar{\zeta}}\Omega_1 + \frac{\bar{\lambda}}{2\bar{\zeta}}\Omega_2 = 0, \qquad \partial_{\zeta}\Omega_2 + \frac{\lambda}{2\zeta}\Omega_1 = 0. \tag{2.10}$$

It should be noted that by the definition of Ω_1^0 the first of equations (2.9) can be written in the form

$$\partial_{\bar{z}}\Omega_1^0 - \frac{\lambda}{2\bar{z}}\overline{\Omega_2} = 0. \tag{2.11}$$

2.3 Let us consider the expressions for Ω_1 and Ω_2 as $\lambda \to 0$. This case is of peculiar interest for us since the singular equation (1.1) for $\lambda = 0$ is transformed into the classical inhomogeneous Cauchy–Rieman system

$$\partial_{\bar{z}} w = f(z), \quad z \in G.$$

For Ω_1 passage to the limit with $\lambda \to 0$ can be realized directly in (2.1). Since according to (2.3) and (2.5)

$$\lim_{\lambda\to 0}\Omega_1^0 = -\frac{1}{2},$$

therefore for Ω_1 we get

$$\Omega_1\Big|_{\lambda=0} = \frac{\zeta}{\zeta - z} - \frac{1}{2}. \tag{2.12}$$

Passing to the limit with $\lambda \to 0$ in equality (2.6), in which Ω_2^0 is defined by (2.7) and (2.8), we obtain

$$\Omega_2\Big|_{\lambda=0} = \frac{1}{2}\lim_{\lambda\to 0}\frac{\lambda}{|\lambda|} = \frac{1}{2}\lim_{\lambda\to 0} e^{i\arg\lambda}.$$

As we are not restricted in subordinating the passage to the limit to any additional conditions we can assume $\lim\limits_{\lambda\to 0}(\arg\lambda) = 0$, and then

$$\Omega_2\Big|_{\lambda=0} = \frac{1}{2}. \tag{2.13}$$

Now we can get an expression of the operator S_G for $\lambda = 0$. From (1.11) taking (2.12) and (2.13) into account we have

$$S_G f = T_G f + c,$$

where

$$T_G f = -\frac{1}{\pi} \iint_G \frac{f(\zeta)}{\zeta - z}\, d\xi\, d\eta \qquad \text{and} \qquad c = \frac{i}{\pi} \mathrm{Im} \iint_G \frac{f(\zeta)}{\zeta}\, d\xi\, d\eta.$$

Thus the operator S_G constructed specially for the singular equation (1.1) for $\lambda \neq 0$ in the regular case ($\lambda = 0$) is converted within the functional c into the main two-dimensional integral operator T_G of generalized analytic function theory [69].

So we can treat the systems of basic kernels Ω_1/ζ and $\Omega_2/\bar{\zeta}$ and the operator S_G as a natural development of the Cauchy kernel $1/(\zeta - z)$ and the operator T_G in the case of generalized Cauchy–Riemann systems with a singular point.

3 Properties of a general operator

According to (1.11) this operator is defined by the equality

$$S_G f = -\frac{1}{\pi} \iint_G \left[\frac{\Omega_1(z,\zeta)}{\zeta} f(\zeta) + \frac{\Omega_2(z,\zeta)}{\bar{\zeta}} \overline{f(\zeta)} \right] d\xi\, d\eta. \tag{3.1}$$

We shall suppose that the domain G is a disk, from now on, without further comment.

Let G be a bounded domain of the complex plane E containing $z = 0$ inside it. We denote the completion of the domain G by $\bar{G}$, and the spaces of continuous functions and of functions p–summable in $\bar{G}$ by $C(G)$ and $L_p(\bar{G})$, respectively. For the norm of the element f in these spaces we shall use the notation $\|f\|_C$ and $\|f\|_{L_p}$.

3.1 The study of the properties of the operator S_G is based on two different representations. The first of them,

$$S_G f = -\frac{1}{\pi} \iint_G \frac{f(\zeta)}{\zeta - z}\, d\xi\, d\eta + S_G^0 f, \tag{3.2}$$

$$S_G^0 = -\frac{1}{\pi} \iint_G \left[\frac{\Omega_1^0(z,\zeta)}{\zeta} f(\zeta) + \frac{\Omega_2(z,\zeta)}{\bar{\zeta}} \overline{f(\zeta)} \right] d\xi\, d\eta, \tag{3.3}$$

follows from (3.1) where instead of Ω_1 we use the right side of formula (2.1). The second one

$$S_G f = -\frac{1}{\pi}\iint\limits_G \frac{f(\zeta)}{\zeta - z}\,d\xi\,d\eta - \frac{\lambda}{\pi}\iint\limits_G \frac{1}{\bar\zeta}\ln\frac{1}{|\zeta - z|}\overline{f(\zeta)}d\xi\,d\eta$$
$$-\frac{\lambda}{\pi}\iint\limits_G \frac{1}{\bar\zeta}\ln g(z,\zeta)\overline{f(\zeta)}\,d\xi\,d\eta + S_G^{00}f, \tag{3.4}$$

$$S_G^{00}f = -\frac{1}{\pi}\iint\limits_G \left[\frac{\Omega_1^0(z,\zeta)}{\zeta}f(\zeta) + \frac{\Omega_2^0(z,\zeta)}{\bar\zeta}\overline{f(\zeta)}\right]d\xi\,d\eta,$$

$$g(z,\zeta) = \begin{cases} |\zeta|, & \text{if } |z| \le |\zeta|, \\ |z|, & \text{if } |z| \ge |\zeta|, \end{cases}$$

is extracted from (3.1) if instead of Ω_1, Ω_2 we put (2.1), (2.6).

Lemma 3.1 *Let $f(z) \in L_p(\bar G), p > 2$. Then $h(z) = S_G f$ satisfies the inequality*

$$|h(z)| < NR^{\frac{(p-2)}{p}}\|f\|_{L_p}, \quad z \in E, \tag{3.5}$$

where R *is the maximal distance from $z = 0$ to the boundary of* G *and N is a constant independent of R.*

In fact, we apply (3.2). For the first term on the right the estimate has been obtained in [69], §6, *ch*.1:

$$\left|\iint\limits_G \frac{f(\zeta)}{\zeta - z}\,d\xi\,d\eta\right| \le \left[\frac{2\pi(p-1)}{p-2}\right]^{\frac{(p-1)}{p}} \cdot (2R)^{\frac{(p-2)}{p}} \cdot \|f\|_{L_p}. \tag{3.6}$$

Further, applying the Holder inequality [29] to every term of the right of (3.3), we have

$$|S_G^0 f| \le \frac{1}{\pi}\|f\|_{L_p} \cdot \left(\iint\limits_G |\zeta|^{-r} d\xi\,d\eta\right)^{1/r}$$
$$\times\left\{\left(\iint\limits_G |\Omega_1^0(z,\zeta)|^q - d\xi\,d\eta\right)^{1/q} + \left(\iint\limits_G |\Omega_2(z,\zeta)|^q\,d\xi d\eta\right)^{1/q}\right\} \tag{3.7}$$

where $1/p+1/q+1/r=1, p>2, q>2, 1<r<2$. From [69] it follows that

$$\left(\iint_G |\zeta|^{-r}\,\mathrm{d}\xi\,\mathrm{d}\eta\right)^{1/r} \le \left(\frac{2\pi}{2-r}\right)^{1/r}\cdot\left(2R\right)^{(2-r)/r}.$$

We estimate the integrals of Ω_1^0 and Ω_2. From (2.3) and (2.5) by virtue of the inequality (2.4) we obtain

$$|\Omega_1^0| < 1+\frac{|\lambda|^2}{2}\sum_{k=1}^{\infty}\frac{1}{k^2} = 1+\frac{\pi|\lambda|^2}{12}.$$

Hence it follows that

$$\left(\iint_G |\Omega_1^0(z,\zeta)|^q\,\mathrm{d}\xi\,\mathrm{d}\eta\right)^{1/q} < M_1 R^{2/q}, \tag{3.8}$$

where

$$M_1 = \pi^{1/q}\left(1+\frac{\pi|\lambda|^2}{12}\right).$$

The function Ω_2 has different analytic representations for $|z|<|\zeta|$ and $|z|>|\zeta|$. We take this into consideration, dividing our domain into two parts and applying further the Minkowski

integral inequality [29]:

$$\left(\iint_G |\Omega_2|^q\mathrm{d}\xi\,\mathrm{d}\eta\right)^{1/q} \le \left(\iint_{0\le|\zeta|\le|z|} |\Omega_2|^q\mathrm{d}\xi\,\mathrm{d}\eta\right)^{1/q} + \left(\iint_{|z|\le|\zeta|\le R} |\Omega_2|^q\mathrm{d}\xi\,\mathrm{d}\eta\right)^{1/q}.$$

Substituting here the expression Ω_2 from (1.12) and using again the Holder and Minkowski inequalities, we get

$$\left(\iint_G |\Omega_2|^q\,\mathrm{d}\xi\,\mathrm{d}\eta\right)^{1/q}$$

$$\le \frac{1}{2}\left(\iint_{0\le\rho\le r}\left(\frac{\rho}{r}\right)^{|\lambda|q}\rho\,\mathrm{d}\rho\,\mathrm{d}\gamma\right)^{1/q} + \sum_{k=1}^{\infty}\frac{|\lambda|}{\mu_k}\left(\iint_{0\le\rho\le r}\left(\frac{\rho}{r}\right)^{\mu_k q}\rho\mathrm{d}\,\rho\,\mathrm{d}\gamma\right)^{1/q}$$

$$+\frac{1}{2}\left(\iint_{r\le\rho\le R}\left(\frac{r}{\rho}\right)^{|\lambda|q}\rho\,\mathrm{d}\rho\,\mathrm{d}\gamma\right)^{1/q} + \sum_{k=1}^{\infty}\frac{|\lambda|}{\mu_k}\left(\iint_{r\le\rho\le R}\left(\frac{r}{\rho}\right)^{\mu_k q}\rho\,\mathrm{d}\rho\,\mathrm{d}\gamma\right)^{1/q}$$

$$\leq \frac{1}{2}\left(2\pi r^{-|\lambda|q} \cdot \frac{\rho^{|\lambda|q+2}}{|\lambda|q+2}\Big|_0^r\right)^{1/q} + \sum_{k=1}^{\infty} \frac{|\lambda|}{\mu_k}\left(2\pi r^{-\mu_k q} \cdot \frac{\rho^{\mu_k q+2}}{-\mu_k q+2}\Big|_0^r\right)^{1/q}$$
$$+\frac{1}{2}\left(2\pi r^{|\lambda|q} \cdot \frac{\rho^{-|\lambda|q+2}}{-|\lambda|q+2}\Big|_r^R\right)^{1/q} + \sum_{k=1}^{\infty} \frac{|\lambda|}{\mu_k}\left(2\pi r^{\mu_k q} \cdot \frac{\rho^{-\mu_k q+2}}{-\mu_k q+2}\Big|_r^R\right)^{1/q}.$$

We note, however, that in integrating the third integral we have assumed for simplicity $|\lambda|q > 2$ (for a fixed λ it can be reached choosing a sufficiently big value of q). Hence we have

$$\left(\iint_G |\Omega_2|^q \, d\xi \, d\eta\right)^{1/q} \leq M_2 R^{2/q}, \tag{3.9}$$

where

$$M_2 = (2\pi)^{1/q} \sum_{k=0}^{\infty} \frac{|\lambda|}{\mu_k}\left[(\mu_k q + 2)^{-1/q} + (\mu_k q - 2)^{-1/q}\right].$$

Using the inequalities (3.8) and (3.9), from (3.7) we get

$$|S_G^0 f| \leq N_0 R^{\frac{(p-2)}{p}} \|f\|_{L_p}, \tag{3.10}$$

where

$$N_0 = \frac{1}{\pi}\left(\frac{2\pi}{2-r}\right)^{1/r} \cdot 2^{(2-r)/r}(M_1 + M_2).$$

Now (3.5) follows from (3.6) and (3.10). Lemma 3.1 is proved.

Lemma 3.2 *Let $f(z) \in L_p(G), p > 2$. Then $h(z) = S_G f$ is continuous on the plane E.*

The assertion of this lemma becomes sufficiently obvious if S_G can be expressed by formula (3.4). The first integral in the right of (3.4) is continuous on E, see [69]. The same thing applies for the other three integrals, as they satisfy the conditions guaranteeing their uniform convergence, and hence their continuity at any point z of the E plane, see [47]. Lemma 3.2 is proved.

Lemma 3.3 *The operator S_G is completely continuous from $L_p(\bar{G}), p > 2$, in $C(\bar{G})$.*

In fact, let $\{f(z)\}$ be some set of functions bounded in $L_p(\bar{G}), p > 2$. We check that S_G maps this set in the set of functions $\{h(z)\}$ compact in $C(\bar{G})$.

From inequality (3.6) the uniform boundedness of the set $\{h(z)\}$ in the norm of C space follows. To complete the proof of Lemma 3.3 we must establish the equicontinuity of this set of functions.

We return to (3.4). As was shown in [69] the first integral on the right of (3.4) with a main singularity in the kernel determines the equicontinuous functions in $\bar{G}$. To prove this fact for the second and third integrals is not difficult. As for the fourth integral, the equicontinuity of functions which are defined by it obviously follows from the continuity of $\Omega_1^0(z,\zeta)$ and $\Omega_2^0(z,\zeta)$ in the two variables z and ζ. Lemma 3.3 has been proved.

3.2 In this subsection formulas (1.2) are substantiated.

We denote the class of functions continuous in G except at the point $z = 0$ by $C(G - 0)$.

Lemma 3.4 *Let $f(z) \in L_p(\bar{G}), p > 2$. Then $h(z) = S_G f$ admits a derivative with respect to z continuous in $G - 0$* and *outside G.*

In G the function $h(z)$ is a solution of the equation

$$\partial_{\bar{z}} h - \frac{\lambda}{2\bar{z}} \bar{h} = f(z). \tag{3.11}$$

Outside the domain G it satisfies the equation

$$\partial_{\bar{z}} h - \frac{\lambda}{2\bar{z}} \bar{h} = 0. \tag{3.12}$$

Proof We express $h(z)$ in the form (3.2) and assume that $z \in G$. According to [69] differentiation of the first integral with respect to $\bar{z}$ gives $f(z)$. In differentiating the second integral we use the properties of the functions Ω_1^0, Ω_2 and the rule of differentiation for an integral depending on a parameter, see [47]. Then we get

$$\partial_{\bar{z}} h = f(z) - \frac{1}{\pi} \iint_G \left[\frac{\partial_{\bar{z}} \Omega_1^0(z,\zeta)}{\zeta} f(\zeta) + \frac{\partial_{\bar{z}} \Omega_2(z,\zeta)}{\bar{\zeta}} \overline{f(\zeta)} \right] d\xi \, d\eta.$$

Expressing by formula (2.9) and (2.2) the derivates $\partial_{\bar{z}} \Omega_1^0$ and $\partial_{\bar{z}} \Omega_2$ through the function Ω_1 and Ω_2 we have

$$\partial_{\bar{z}} h = f(z) - \frac{\lambda}{2\pi \bar{z}} \iint_G \left[\frac{\overline{\Omega_2(z,\zeta)}}{\zeta} f(\zeta) + \frac{\overline{\Omega_1(z,\zeta)}}{\bar{\zeta}} \overline{f(\zeta)} \right] d\xi \, d\eta.$$

Finally, using formula (3.1) for the determination of the operator S_G we come to (3.11).

Now we consider the case $z \notin G$. The first integral in (3.2) outside G is a holomorphic function with respect to z. Therefore its differentiation with respect to $\bar{z}$ gives 0 now, and as a result we get (3.12).

Continuous differentaition of $h(z)$ with respect to $\bar{z}$ in $G-0$ and outside G follows from (3.11), (3.12) by virtue of the properties of $h(z)$ and $f(z)$. Lemma 3.4 is proved.

Let $C_{\bar{z}}(G-0)$ be the class of functions continuously differentiable with respect to $\bar{z}$ in $G-0$. We return again to equations (1.1) and (1.3) assuming that $w(z)$ and $\Phi(z)$ are considered to be in the class $C(\bar{G})\cap C_{\bar{z}}(G-0)$ and $f(z)\in C(G-0)\cap L_q(\bar{G}), q>2$. By virtue of Lemma 3.4 the function $h(z)=S_G f, z\in G$, is a particular solution of equation (1.1); therefore formula (1.2) defines the general solution of this equation from $C(\bar{G})\cap C_{\bar{z}}(G-0)$.

4 The general integral equation

We pass over to the generalized Cauchy–Riemann system with a singular point in the form

$$\partial_{\bar{z}} w - \frac{b(z)}{2\bar{z}}\bar{w} = F(z), \quad z\in G. \tag{4.1}$$

Let $w(z)\in C(\bar{G})\cap C_{\bar{z}}(G-0), F(z)\in C(G-0)\cap L_q(\bar{G}), q>2$, and the coefficient $b(z)$ be continuous in $G, b(0)\neq 0$ and

$$|b(z)-b(0)|<M|z|^{\alpha}, \quad \alpha>0,$$

for z belonging to an arbitrarily small neighbourhood of $z=0$.

We write down (4.1) in the following form:

$$\partial_{\bar{z}} w - \frac{b(0)}{2\bar{z}}\bar{w} = F(z) + \frac{b(z)-b(0)}{2\bar{z}}\bar{w}.$$

If we assume $\lambda=b(0)$ and $f(z)=F(z)+[b(z)-b(0)]/2\bar{z}$ then (4.1) takes the form of equation (1.1), where $f(z)\in C(G-0)\cap L_p(\bar{G}), p=\min(q,2/1-\alpha)$. Therefore formula (1.2) applied to equation (4.1) leads to the following integral equation for $w(z)$:

$$w(z)=\Phi(z)+S_G F+P_G\bar{w}, \quad z\in G, \tag{4.2}$$

$$P_G\bar{w}=S_G\left(\frac{b(z)-b(0)}{2\bar{z}}\overline{w(z)}\right), \tag{4.3}$$

where $\Phi(z)\in C(\bar{G})\cap C_{\bar{z}}(G-0)$ is a solution of equation (1.3).

The integral equation (4.2) is used later as a general apparatus for studying the generalized Cauchy–Riemann system (4.1) with a singular point. The first obvious consequence from (4.2) is:

Theorem 4.1 *The integral equation (4.2) establishes a connection between the sets of continuous solutions of the equations (4.1) and (1.3) in the closed domain G. Namely, if $w(z)$ is a given solution of equation (4.1) then the solution $\Phi(z)$ of equation (1.3), which corresponds to it, is uniquely determined by (4.2). And conversely, if for some continuous solution $\Phi(z)$ of (1.3) the integral solution (4.2) has a continuous solution $w(z)$ then it will satisfy (4.1).*

We shall indicate some properties of the operator P_G. It follows from Lemma 3.3 that P_G is completely continuous in $C(\bar{G})$ and maps this space in itself. From (3.5) the estimate for its norm follows:

$$\|P_G\|_C < NR^{(p-2)/p}\|\frac{b(z)-b(0)}{2\bar{z}}\|_{L_p}, \quad p > 2. \tag{4.4}$$

Hence it is clear that P_G will be a contraction operator, i.e. $\|P_G\|_C < 1$, either because of the smallness of the measure on the domain (R is small) or at the expense of a small difference between $b(z)$ and $b(0)$.

In this condition equation (4.2) has a unique solution and

$$w(z) = \Phi(z) + R_G\Phi + S_GF + R_G(S_GF),$$

where R_G is a completely continuous operator in $C(\bar{G})$, the resolvent of P_G. Thus, the unique continuous $w(z)$ solution of equation (4.1) corresponds to every continuous $\Phi(z)$ solution of equation (1.3).

If we add to this result the assertion of Theorem 4.1, then the following is obvious.

Consequence 4.1 *For $\|P_G\|_C < 1$, between the sets of continuous solutions of equations (4.1) and (1.3) there exists a one–to–one correspondence.*

This shows a natural way of studying properties of the solution of equation (4.1). Firstly, it is necessary to analyse the equation we take an interest in, on a model, i.e. on an example of a simpler equation (1.3), and then to transform the results obtained to the general case by integral equation (4.2).

This scheme is evidently not a discovery. As a rule, it can be found in many various problems. It is widely and obviously represented in studying regular generalized Cauchy–Riemann systems [69], with respect to which classical generalized Cauchy–Riemann systems defining analytic functions of a complex variable are used as a model. The above scheme suggests that an appropriate theory of a model equation is either available or wants preliminary elaboration. For the given singular equation

(4.1) we prove to be in the second situation. Therefore the whole of the following chapter will be entirely devoted to constructing a theory of the model equation.

5 Additions to Chapter 1

In papers by N. Bliev and A. Tungatarov [14, 50, 51] the basic S_G operator is considered in Besov's fractional spaces $\mathcal{B}^{\alpha}_{p,1}(G)$ on condition that $\alpha = (2/p) - 1$ and $1 < p < 2$, [9]. In this case the domain of definition of the operator is wider than in §3, because elements of the above spaces composing a subset of $L_2(\bar{G})$, generally speaking, do not belong to $L_q(\bar{G}), q > 2$. The operator S_G is characterized by the following properties.

Theorem 5.1 *Let $f(z) \in \mathcal{B}^{\alpha}_{p,1}(G)$, where $\alpha = (2/p) - 1$ and $1 < p < 2$. Then the function $h(z) = S_G f$ is continuous and bounded in the whole complex plane E, and*

$$\|S_G f\|_{C(\bar{G})} \leq N \|f\|_{L_2(\bar{G})},$$

where N is a constant.

Theorem 5.2 *In the statement of Theorem 5.1 the function $h(z)$ admits Sobolev's generalized derivative with respect to $\bar{z}$ which belongs to $L_p(G), 1 < p < 2$, and satisfies almost everywhere equation (1.1) in G.*

These theorems comprise a base for studying equation (4.1) assuming that $b(z)$ and $F(z)$ belong to the function space $\mathcal{B}^{\alpha}_{p,1}(G)$; see [51] for details.

6 Unsolved problems

6.1 In §4 the solvability of the main integral equation

$$w(z) = \Phi(z) + S_G F + P_G \bar{w}, \quad z \in G, \tag{6.1}$$

was formulated only by fulfilling the conditions $\|P_G\|_C < 1$. This provides a one–to–one correspondence between the sets $\{w(z)\}$ and $\{\Phi(z)\}$ of the general and model equation solutions with a singular point. If we ignore this restriction, then we must be satisfied with a more moderate achievement: a correspondence between $\{w(z)\}$ and $\{\Phi(z)\}$ is established by the completely continuous operator P_G.

Is this result completed or can it be detailed, having proved the one–valued solvability $\{w(z)\}$ through $\{\Phi(z)\}$?

6.2 The theory of generalized analytic functions prompts one of several possible ways in solving the question formulated above. The proof of its main statements is known to rely on two integral representations which connect generalized analytic functions with analytic functions of a complex variable [69]. The representation of the first type, called a *formula of reciprocity* or a *formula of similarity*, is used for determining a manifold of generalized analytic functions on all of the complex plane E (see Theorems 3.2 and 3.2′ from [69]). This result, in turn, proves to be important for studying the question of solvability of the basic integral equation, that is the integral representation of the second type.

If we want to use this scheme for studying the solvability of equation (6.1) then we should turn to 'its' own formula of similarity

$$w(z) = \Phi(z)e^{\omega(z)}, \quad z \in G, \tag{6.2}$$

in which $w(z), \Phi(z)$ and $\omega(z)$ satisfy in G the equations

$$\partial_{\bar{z}} w - \frac{b(z)}{2\bar{z}} \bar{w} - 0, \tag{6.3}$$

$$\partial_{\bar{z}} \Phi - \frac{b(0)}{2\bar{z}} \bar{\Phi} = 0, \tag{6.4}$$

$$\partial_{\bar{z}} \omega + \frac{b(0)}{2\bar{z}} \frac{\bar{w}}{w} e^{2i \operatorname{Im} \omega} = \frac{b(z)\bar{w}}{w}. \tag{6.5}$$

The relation (6.5) is obtained from (6.2) by differentiating with respect to $\bar{z}$ and excluding $\partial_{\bar{z}} w$ and $\partial_{\bar{z}} \Phi$ by means of (6.3) and (6.4). This relation proves to be more complicated in comparison to the same one for generalized analytic functions.

Up to now the question of *whether it is possible for every solution* $w(z) \in C(\bar{G}) \cap C_{\bar{z}}(G - 0)$ *of equation (6.3) to search a continuous solution of equation (6.5) in* G*, remains open.*

If this were proved, then formula (6.2) would be transformed into an effective tool for investigating the solvability of the basic integral equation (6.1).

Chapter 2
The model equation

In this chapter the object of study is the equation

$$\partial_{\bar z}\Phi - \frac{\lambda}{2\bar z}\bar\Phi = 0, \quad z \in G.$$

where λ is a complex number and G is a domain containing the point $z = 0$ inside itself. Our direct aim is constructing a base of the theory for solutions of this equation like the theory of analytic functions of a complex variable.

The results of this chapter have been published in [52, 57].

1 Basic kernels and elementary solutions of the conjugate equation

The study of the model equation is closely connected with the conjugate equation

$$\partial_{\bar z}\Phi^* + \frac{\bar\lambda}{2z}\bar\Phi^* = 0, \quad z \in G. \tag{1.1}$$

In the present section, according to the scheme described in I. Vekua's monograph [69], on the basis of the known general kernels of the model equation we construct the general kernels and elementary solutions of equation (1.1) and formulate their properties which will be necessary for future use.

1.1 We shall call the functions $\Omega_1^*(z,\zeta)/\zeta$ and $\Omega_2^*(z,\zeta)/\zeta$ the *general kernels* of equation (1.1) and define them according to the formula

$$\frac{\Omega_1^*(z,\zeta)}{\zeta} = -\frac{\Omega_1(\zeta,z)}{z}, \qquad \frac{\Omega_2^*(z,\zeta)}{\bar\zeta} = -\frac{\overline{\Omega_2(\zeta,z)}}{z}, \tag{1.2}$$

through the known functions $\Omega_1(z,\zeta)$ and $\Omega_2(z,\zeta)$ obtained in §1 ch. 1.

From (2.10) of ch.1 it follows that

$$\partial_{\bar z}\Omega_1^* - \frac{\bar\lambda}{2z}\overline{\Omega_2^*} = 0, \qquad \partial_{\bar z}\Omega_2^* + \frac{\bar\lambda}{2z}\,\overline{\Omega_1^*} = 0. \tag{1.3}$$

From Lemmas 2.1 and 2.2 of ch.1 we have the following properties of the kernels

$$\begin{aligned} \frac{1}{\zeta}\,\Omega_1^*(z,\zeta) &= \frac{1}{\zeta - z} + \tilde{\Omega}_1^*(z,\zeta) \\ \frac{1}{\bar{\zeta}}\,\Omega_2^*(z,\zeta) &= -\frac{\lambda}{z}\ln\frac{1}{|\zeta - z|} + \tilde{\Omega}_2^*(z,\zeta), \end{aligned} \tag{1.4}$$

where $\tilde{\Omega}_1^*$ and $\tilde{\Omega}_2^*$ are continuous with respect to $z, \zeta \in E$ everywhere except for $z = 0$ and $z = \zeta = \infty$.

For a fixed value of $z \neq 0, \infty$ the functions Ω_1^*/ζ and $\Omega_2^*/\bar{\zeta}$ with respect to ζ at the points $\zeta = 0$ and $\zeta = \infty$ have a zero of order $|\lambda|$. If a value $\zeta \neq 0, \infty$ is fixed then the above functions for $z = \infty$ have a zero of order $1 + |\lambda|$, and for $z = 0$ have a pole of order $1 - |\lambda|$ as $|\lambda| < 1$, and a zero of order $|\lambda| - 1$ as $|\lambda| > 1$.

1.2 We take into consideration the two new functions $X_1^*(z,\zeta)$ and $X_2^*(z,\zeta)$:

$$\begin{aligned} X_1^*(z,\zeta) &= \frac{1}{2}\left(\frac{\Omega_1^*(z,\zeta)}{\zeta} + \frac{\Omega_2^*(z,\zeta)}{\bar{\zeta}}\right), \\ X_2^*(z,\zeta) &= \frac{1}{2\mathrm{i}}\left(\frac{\Omega_1^*(z,\zeta)}{\zeta} - \frac{\Omega_2^*(z,\zeta)}{\bar{\zeta}}\right). \end{aligned} \tag{1.5}$$

According to the previous subsection these functions can be expressed in the form

$$X_1^*(z,\zeta) = \frac{1}{2(\zeta - z)} - \frac{\bar{\lambda}}{2z}\ln\frac{1}{|\zeta - z|} + \tilde{X}_1^*(z,\zeta),$$

$$X_2^*(z,\zeta) = \frac{1}{2\mathrm{i}(\zeta - z)} - \frac{\bar{\lambda}}{2\mathrm{i}z}\ln\frac{1}{|\zeta - z|} + \tilde{X}_2^*(z,\zeta),$$

where $\tilde{X}_1^*$ and $\tilde{X}_2^*$ are continuous with respect to z and ζ everywhere except for $z = 0$ and $z = \zeta = \infty$. The behaviour of these functions with respect to ζ at the points $\zeta = 0$ and $\zeta = \infty$, when the value $z \neq 0, \infty$ is fixed, and with respect to z at the points $z = 0$ and $z = \infty$, when the value $\zeta \neq 0, \infty$ is fixed, is similar to the behaviour of the general kernels Ω_1^*/ζ and $\Omega_2^*/\bar{\zeta}$ of the conjugate equation.

According to (1.5) and (1.3) the functions X_k^*, $k = 1, 2$, satisfy equation (1.1):

$$\partial_{\bar{z}} X_k^* + \frac{\bar{\lambda}}{2z}\overline{X_k^*} = 0, \quad z \in G.$$

These functions are called *elementary solutions* of equation (1.1).

2 Cauchy generalized formula

A fundamental part of the construction of a theory of the model equation belongs to the Cauchy generalized formula, i.e. to a relation admitting the calculation of the solution to this equation at any point of the domain if its values on a boundary are known. In this section we prove this formula using Vekua's scheme which is based on Green's identity and on general kernels and elementary solutions of equation (1.1) [69].

2.1 Simultaneously with the model equation

$$\partial_{\bar{z}}\Phi - \frac{\lambda}{2\bar{z}}\bar{\Phi} = 0, \quad z \in G. \tag{2.1}$$

we consider the conjugate one

$$\partial_{\bar{z}}\Phi^* + \frac{\bar{\lambda}}{2z}\overline{\Phi^*} = 0, \quad z \in G. \tag{2.2}$$

Let G be a multiply–connected domain bounded with a finite number of simple smooth closed contours Γ and containing the point $z = 0$ inside itself. If $\Phi(z)$ and $\Phi^*(z)$ are two arbitrary solutions of equations (2.1) and (2.2) from the class $C(G+\Gamma) \cap C_{\bar{z}}(G-0)$, then they satisfy the relation

$$\mathrm{Re}\left(\frac{1}{2\mathrm{i}}\int_\Gamma \Phi(z)\Phi^*(z)\,\mathrm{d}z\right) = 0, \tag{2.3}$$

which is called Green's identity and expresses the property of self–conjugacy for equations (2.1) and (2.2). The proof of identity (2.3), described in Vekua's monograph ([69], §9, ch. 3) in the case of regular Cauchy–Rieman systems, is extended without any difficulties into solutions of equations (2.1) and (2.2).

2.2 We pass directly to deducing the Cauchy generalized formula. For this purpose, as $\Phi^*(z)$ we take the elementary solutions $X_k^*(z,t), k = 1,2$, of equation (2.2). Let us first suppose $t \in G$ and denote the circle $|z-t| = \varepsilon$, where ε is a small positive number, by Γ_ε. Using formula (2.3) in the domain, bounded with Γ and Γ_ε, we get

$$\int_\Gamma X_k^*(z,t)\Phi(z)\,\mathrm{d}z - \overline{X_k^*(z,t)\Phi(z)}\,\overline{\mathrm{d}z}$$

$$= \int_{\Gamma_\varepsilon} X_k^*(z,t)\Phi(z)\,\mathrm{d}z - \overline{X_k^*(z,t)\Phi(z)}\,\overline{\mathrm{d}z} \quad (k = 1,2).$$

Multiplying the second equality ($k = 2$) by i and adding it to the first one ($k = 1$) by virtue of (1.5), we get

$$\int_{\Gamma} \frac{\Omega_1^*(z,t)}{t}\Phi(z)\,\mathrm{d}z - \overline{\frac{\Omega_2^*(z,t)}{t}}\,\overline{\Phi(z)}\,\overline{dz} = \int_{\Gamma_\varepsilon} \frac{\Omega_1^*(z,t)}{t}\Phi(z)\,\mathrm{d}z - \overline{\frac{\Omega_2^*(z,t)}{t}}\,\overline{\Phi(z)}\,\overline{\mathrm{d}z}.$$

Passing over to the limit with $\varepsilon \to 0$ and taking (1.4) into account we get

$$\int_{\Gamma} \frac{\Omega_1^*(z,t)}{t}\Phi(z)\,\mathrm{d}z - \overline{\frac{\Omega_2^*(z,t)}{t}}\,\overline{\Phi(z)\mathrm{d}z} = -2\pi \mathrm{i}\Phi(t).$$

In a similar way the variants $t \in \Gamma$ and $t \notin G + \Gamma$ are analysed without dwelling on it, we write out the final formula at once:

$$-\frac{1}{2\pi \mathrm{i}}\int_{\Gamma} \frac{\Omega_1^*(\zeta,z)}{z}\Phi(\zeta)\,\mathrm{d}\zeta - \overline{\frac{\Omega_2^*(\zeta,z)}{z}}\,\overline{\Phi(\zeta)}\,\overline{\mathrm{d}\zeta} = \begin{cases} \Phi(z) & \text{for } z \in G, \\ \frac{\alpha}{2}\Phi(z) & \text{for } z \in \Gamma, \\ 0 & \text{for } z \notin G+\Gamma, \end{cases} \tag{2.4}$$

where $\alpha\pi$ denotes the interior angle of the contour Γ at z.

If in (2.4) the general kernels of the conjugate equation are expressed by (1.2) through the general kernels of the model equation (2.1), then (2.4) will take the form

$$\frac{1}{2\pi \mathrm{i}}\int_{\Gamma} \frac{\Omega_1(z,\zeta)}{\zeta}\Phi(\zeta)\,\mathrm{d}\zeta - \frac{\Omega_2(z,\zeta)}{\bar\zeta}\,\overline{\Phi(\zeta)\,\mathrm{d}\zeta} = \begin{cases} \Phi(z) & \text{for } z \in G, \\ \frac{\alpha}{2}\Phi(z) & \text{for } z \in \Gamma, \\ 0 & \text{for } z \notin G+\Gamma. \end{cases} \tag{2.5}$$

We call it the *generalized Cauchy formula* for equation (2.1).

Remark Let $w(z)$ be a solution of the inhomogeneous equation

$$\partial_{\bar z} w - \frac{\lambda}{2\bar z}\bar w = f(z), \quad z \in G. \tag{2.6}$$

where $f(z) \in L_p(G+\Gamma), p > 2$. As in [69] (§10, ch. 3) it is easy to obtain the relation:

$$\frac{1}{2\pi \mathrm{i}}\int_{\Gamma} \frac{\Omega_1(z,\zeta)}{\zeta}w(\zeta)\,\mathrm{d}\zeta - \frac{\Omega_2(z,\zeta)}{\bar\zeta}\,\overline{w(\zeta)\,\mathrm{d}\zeta} \tag{2.7}$$

$$-\frac{1}{\pi}\iint\limits_G \left[\frac{\Omega_1(z,\zeta)}{\zeta}f(\zeta) + \frac{\Omega_2(z,\zeta)}{\bar\zeta}\overline{f(\zeta)}\right]\mathrm{d}\xi\,\mathrm{d}\eta = \begin{cases} w(z) & \text{for } z \in G, \\ \frac{\alpha}{2}w(z) & z \in \Gamma, \\ 0 & \text{for } z \notin G+\Gamma. \end{cases}$$

3 Sequences of continuous solutions for the model equation

The uniform convergence of a sequence of holomorphic functions is known to guarantee the holomorphy of its limit function; see for example [48]. A similar situation takes place for solutions of the model equation. Namely, the following theorem is valid.

Theorem 3.1 *Let $\Phi_n(z), n = 1, 2, \ldots$, be a sequence of solutions of equation (2.1) which belong to $C(G) \cap C_{\bar{z}}(G - 0)$. If this sequence is uniformly convergent inside G then the limit function*

$$\Phi(z) = \lim_{n\to\infty} \Phi_n(z)$$

is also a solution from the class $C(G) \cap C_{\bar{z}}(G - 0)$ of equation (2.1).

Proof First of all, it should be noted that by the uniform convergence in G we understand the uniform convergence in any closed bounded subdomain contained in G. It is quite clear that we should take only such subdomains Δ into consideration where $z = 0$ is an interior point of Δ.

Let $\partial\Delta$ be the boundary of $\Delta(\Delta + \partial\Delta \subset G)$. Then thanks to the generalized Cauchy formula (2.5) for any n and $z \in \Delta$ we have

$$\Phi_n(z) = \frac{1}{2\pi i}\int_{\partial\Delta} \frac{\Omega_1(z,\zeta)}{\zeta}\Phi_n(\zeta)\,d\zeta - \frac{\Omega_2(z,\zeta)}{\bar{\zeta}}\overline{\Phi_n(\zeta)\,d\zeta}.$$

As the sequences $\Omega_1\Phi_n/\zeta$ and $\Omega_2\bar{\Phi}_n/\bar{\zeta}$ converge uniformly with respect to ζ on the contour $\partial\Delta$ then passing to the limit in the two parts of the equality we get

$$\Phi(z) = \frac{1}{2\pi i}\int_{\partial\Delta} \frac{\Omega_1(z,\zeta)}{\zeta}\Phi(\zeta)\,d\zeta - \frac{\Omega_2(z,\zeta)}{\bar{\zeta}}\overline{\Phi(\zeta)}d\bar{\zeta}. \tag{3.1}$$

Another formulation of this result is

Theorem 3.2 *Let $\Phi_n(z), n = 1, 2, \ldots$, be a sequence of solutions of equation (2.1) from the class $C(G + \Gamma) \cap C_{\bar{z}}(G - 0)$, uniformly converging on Γ. Then $\lim_{n\to\infty} \Phi_n(z)$ exists in G and defines a solution of equation (2.1) from the same class.*

4 $\Phi(z)$ representation by some series. Analogy of the Liouville theorem. Uniqueness theorem

4.1 We denote the disk $|z| < R$ by K and its boundary $|z| = R$ by ∂K. Let us use the representation of the solution $\Phi(z)$ of (2.1) by means of the generalized

Cauchy integral (3.1). Substituting $\zeta = \mathrm{Re}^{i\theta}$ into (3.1) and applying the notation $\Phi(\theta) = \Phi(\mathrm{Re}^{i\theta})$, we get

$$\Phi(\theta) = \frac{1}{2\pi}\int_0^{2\pi}\left\{\Omega_1(z, \mathrm{Re}^{i\theta})\Phi(\theta) + \Omega_2(z, \mathrm{Re}^{i\theta})\overline{\Phi(\theta)}\right\} d\theta.$$

We turn to the concrete forms of the kernels Ω_1 and Ω_2, see (1.12) of ch. 1. Introducing their expressions (which correspond to the inequality $|z| < |\zeta| = R$) under the integral symbol and realizing a term–by–term integration, we have

$$\Phi(z) = (\alpha_0 + \frac{\lambda}{|\lambda|}\overline{\alpha_0})r^{|\lambda|} + \sum_{k=1}^{\infty}\left\{\frac{P_k\alpha_k + \lambda\bar{a}_{-k}}{\mu_k}e^{ik\varphi} + \frac{P_{-k}a_{-k} + \lambda\bar{a}_k}{\mu_k}e^{-ik\varphi}\right\}r^{\mu_k} \quad (4.1)$$

where

$$a_k = \frac{R^{-\mu_k}}{4\pi}\int_0^{2\pi}\Phi(\theta)e^{-ik\theta}d\theta, \quad k = 0, \pm 1, \ldots \quad (4.2)$$

To provide the absolute and uniform convergence of series (4.1) for $|z| < R$, it is sufficient to subordinate $\Phi(z)$ to the absolute integrability condition on ∂K. It is easy to see that this condition also guarantees $\Phi(z)$ infinite differentiability with respect to z and $\bar{z}$ for $z \in K - 0$.

The following result is an obvious consequence of the representation of $\Phi(z)$ in the form of the series (4.1).

Theorem 4.1 *Any solution of the model equation from the class $C(G+\Gamma)\cap C_{\bar{z}}(G-0)$ vanishes at the point $z = 0$. Possible orders of the zero are $\mu_k = \sqrt{|\lambda|^2 + |k|^2}, k = 0, 1, \ldots$.*

4.2 From equation (4.2) an estimate for the coefficient of series (4.1) is obtained:

$$|a_k| < \frac{M(R)}{2R^{\mu_k}}, \quad k = 0, \pm 1, \ldots, \quad (4.3)$$

where $M(R) = \max|\Phi(z)|$ on the circle $|z| = R$.

On the basis of (4.3) we can prove

Theorem 4.2 **(Analogy of Liouville's Theorem)** *If a solution of the model equation from the class $C(E)\cap C_{\bar{z}}\,(E-0)$ is bounded on the whole complex plane E, then $\Phi(z) \equiv 0$.*

In fact, inequality (4.3) with regard to the boundedness $|\Phi(z)| < M$, where $z \in E$, takes the form

$$|a_k| < \frac{M}{2R^{\mu_k}}, \quad k = 0, \pm 1, \ldots$$

Since this is valid for any value of R then letting R tend to infinity we get

$$|a_k| = 0, \quad \text{i.e.} \quad a_k = 0 \quad k = 0, \pm 1, \ldots$$

Now from (4.1) it follows that $\Phi(z) \equiv 0$.

We now pay attention to a specific form of Theorem 4.2. In contrast to a similar theorem for analytic functions, among which there are functions bounded on the whole plane, such functions are absent from the solutions of the model equation.

4.3 In this subsection we prove a uniqueness theorem for solutions of equation (2.1).

Theorem 4.3 *If a solution of the model equation from the class $C(G+\Gamma) \cap C_{\bar{z}}(G-0)$ vanishes at some infinite set of points which have a limit point, belonging to the domain G, then $\Phi(z) \equiv 0$ in G.*

In fact, if a limit point z_0 does not coincide with $z = 0$ then the theorem is a consequence of the uniqueness theorem for generalized analytic functions, see [69] §4 ch.3. If $z_0 = 0$, then to prove the theorem we must return to relation (4.1), assuming that it is written out for some disk K of the point $z_0 = 0$ which is entirely contained in the domain G.

First we prove that $\Phi(z) \equiv 0$ in K. Our reasoning will be developed according to the scheme applied to proving the uniqueness theorem for the case of analytic functions, see for example [34] p. 174. To this end, from the set of points in which $\Phi(z)$ vanishes according to the conditions of the theorem, we select a sequence of different points $z_1, z_2, \ldots, z_p, \ldots$, convergent to $z_0 = 0$. From (4.1) with regard to the notation $z_p = r_p e^{i\varphi_p}$ we get

$$\begin{aligned} \Phi(z_p) =& (a_0 + \frac{\lambda}{|\lambda|}\overline{a_0}) r_p^{|\lambda|} \\ &+ \sum_{k=1}^{\infty} \left\{ \frac{P_k a_k + \lambda \overline{a_{-k}}}{\mu_k} e^{ik\varphi_p} + \frac{P_{-k} a_{-k} + \lambda \bar{a}_k}{\mu_k} e^{-ik\varphi_p} \right\} r_p^{\mu_k} = 0. \end{aligned} \tag{4.4}$$

Dividing this equality by $r_p^{|\lambda|}$ and then passing over to the limit as $z_p \to 0$, we have

$$a_0 + \frac{\lambda}{|\lambda|}\overline{a_0} = 0.$$

Therefore relation (4.4) has the form:

$$\sum_{k=1}^{\infty}\left\{\frac{P_k a_k + \lambda\overline{a_{-k}}}{\mu_k}e^{ik\varphi_p} + \frac{P_{-k}a_{-k} + \lambda\overline{a_k}}{\mu_k}e^{-ik\varphi_p}\right\}r_p^{\mu_k} = 0.$$

Once again, dividing this by $r_p^{\mu_1}$ and passing over to the limit with $z_p \to 0$, we obtain

$$P_1 a_1 + \lambda\overline{a_{-1}} = 0, \quad \lambda\overline{a_1} + P_{-1}a_{-1} = 0.$$

Fulfilling the similar operations, step by step, for any index $k = 2, 3, \dots$, we get

$$P_k a_k + \lambda\overline{a_{-k}} = 0, \qquad \lambda\overline{a_k} + P_{-k}a_{-k} = 0.$$

Thus, all the coefficients of series (4.1) are equal to zero, but in this case $\Phi(z) \equiv 0$ in K. As to the other part of G, viz. for $G - K, \Phi(z) \equiv 0$ as well. This follows from the uniqueness theorem for generalized analytic functions. Theorem 4.3 has been proved.

4.4 The representation of the solution $\Phi(z)$ for the model equation in the form of the power series (4.1) is sometimes conveniently used in the different form

$$\Phi(z) = \mathcal{X} A_0 r^{|\lambda|} + \sum_{k=1}^{\infty}\left(\lambda A_k e^{ik\varphi} + P_{-k}\bar{A}_k e^{-ik\varphi}\right)r^{\mu_k}, \tag{4.5}$$

in which A_0 is a real constant and A_k is a complex constant, $\mathcal{X} = 1$ for $\lambda > 0$; $\mathcal{X} = \mathrm{i}$ for $\lambda < 0$; and $\mathcal{X} = \mathrm{i}(|\lambda| - \lambda)$ for Im $(\lambda) \neq 0$.

The equivalence of formulas (4.1) and (4.5) is proved in the following way. According to simple calculations the coefficients in (4.1) satisfy the equalities

$$\mathrm{Im}\frac{1}{\mathcal{X}}(a_0 + \frac{\lambda}{|\lambda|}\overline{a_0}) = 0 \qquad P_{-k}\frac{P_k a_k + \lambda\overline{a_{-k}}}{\mu_k} = \lambda\overline{\left\{\frac{P_{-k}a_{-k} + \lambda\overline{a_k}}{\mu_k}\right\}}.$$

Hence it follows that if in place of a_0, a_k and a_{-k} we use the constants A_0 (real) and A_k defined by

$$A_0 = \frac{1}{\mathcal{X}}(a_0 + \frac{\lambda}{|\lambda|}\overline{a_0}), \qquad \lambda A_k = \frac{1}{\mu_k}(P_k a_k + \lambda\bar{a}_{-k}),$$

then relation (4.1) will take the form (4.5). Replacing a_0 and a_k by means of (4.2) in these formulas, we get

$$\begin{aligned} A_0 =& \frac{R^{-|\lambda|}}{4\pi\mathcal{X}} \int_0^{2\pi} \left(\Phi(\theta) + \frac{\lambda}{|\lambda|} \overline{\Phi(\theta)} \right) d\theta, \\ A_k =& \frac{R^{-\mu_k}}{4\pi\mathcal{X}} \int_0^{2\pi} \left(\frac{P_k}{\lambda} \Phi(\theta) + \overline{\Phi(\theta)} \right) e^{-ik\theta} d\theta. \end{aligned} \tag{4.6}$$

These equalities determine A_0 and A_k through values of the solution $\Phi(z)$ on the boundary $|z| = R$.

5 Regularity of solutions at a singular point

An arbitrary solution of the model equation from the class $C(G) \cap C_{\bar{z}}(G-0)$ is an analytic function of the variables z and $\bar{z}$ outside the point $z = 0$ because of the properties of elliptic systems. But the analyticity of the solution at the singular point is not generally guaranteed. Moreover, formula (4.1) implies that different solutions can belong to various classes of regularity at this point. In the present section we consider this problem.

5.1 We suppose that the model equation

$$2\bar{z}\partial_{\bar{z}}\Phi - \lambda\bar{\Phi} = 0 \tag{5.1}$$

in a neighbourhood of the point $z = 0$ admits an analytic solution

$$\Phi = \sum_{k=1}^{\infty} \Phi_m(z, \bar{z}), \quad \text{where} \quad \Phi_m(z, \bar{z}) = \sum_{s=0}^{m} a_{s,m-s} z^s \bar{z}^{(m-s)}.$$

In this case because of the specific equation (5.1) each of the homogeneous forms Φ_m is a separate solution of this equation. Let us determine the condition which guarantees that at least the coefficient $a_{s,m-s}$ will not be equal to zero, and therefore the equation will allow a non–trivial analytic solution.

Assuming $z = re^{i\varphi}$, we rewrite Φ_m in the form

$$\Phi_m(z, \bar{z}) = r^m f_m(\varphi), \tag{5.2}$$

where

$$f_m(\varphi) = \sum_{s=0}^{m} a_{s,m-s} e^{i(2s-m)\varphi}. \tag{5.3}$$

Substitution of (5.2) into (5.1) gives

$$\mathrm{i}\frac{\mathrm{d}f_m}{\mathrm{d}\varphi} + mf_m - \lambda\bar{f}_m = 0. \tag{5.4}$$

For $m = 0$ from (5.3) we have $f_0(\varphi) = a_{0,0}$, and from (5.4) it follows that $a_{O,O} = 0$. So we believe that $m \geq 1$.

Integrating (5.4) we have

$$f_m(\varphi) = \lambda c_m \mathrm{e}^{\mathrm{i}\sqrt{m^2-|\lambda|^2}\varphi} + \bar{c}_m\left(m - \sqrt{m^2-|\lambda|^2}\right)\mathrm{e}^{-\mathrm{i}\sqrt{m^2-|\lambda|^2}\varphi} \tag{5.5}$$

where c_m is a complex constant. The function $f_m(\varphi)$ as is seen from (5.3), must be periodic with period 2π. This means that $\sqrt{m^2 - |\lambda|^2} = k$, where k is a positive integer. Further eliminating m from (5.5) and (5.2) by this equality and substituting simultaneously (5.5) into (5.2), we get

$$\Phi_m(z, \bar{z}) = (\lambda C_k \mathrm{e}^{\mathrm{i}k\varphi} + P_{-k}\bar{C}_k \mathrm{e}^{-\mathrm{i}k\varphi})r^{\mu_k}. \tag{5.6}$$

Here we applied the usual notation $\mu_k^2 = |\lambda|^2 + k^2, P_{-k} = \mu_k - k$, and introduced the new constant $C_k = c_m, m = \sqrt{|\lambda|^2 + k^2}$.

The right side of equality (5.6) coincides with the k term in formula (4.5). But this means that equation (5.1) allows an analytic solution with respect to z and $\bar{z}$ if and only if at least one of the terms in (4.5) is an analytic function.

For this purpose it is necessary and sufficient that the equality

$$\mu_k = k + 2p \tag{5.7}$$

be realized for some $k \geq 0$ and a natural number p. Resolving equality (5.7) with respect to $|\lambda|$, we get

$$|\lambda|^2 = 4p(p + k). \tag{5.8}$$

Thus, we come to the following result.

Theorem 5.1 *Only for values of* λ *which belong to the set* $\Lambda_0 = \{\lambda : |\lambda|^2 = 4p(p+k), k \geq 0, p \geq 1\}$ *does equation (5.1) have a solution analytic in the variables* z *and* $\bar{z}$ *in a neighbourhood of the singular point. For other* λ this type of solution does not exist.

5.2 Let us introduce the auxiliary plane Λ from which the points $\lambda = 0$ and $\lambda = \infty$ are excluded. We denote by Λ_1 the complement of the set Λ_0 with respect to $\Lambda\,(\Lambda_1 = \Lambda - \Lambda_0)$ and by V a circular neighbourhood of the point $z = 0$. We prove the following assertion.

Theorem 5.2 *Let $\lambda \in \Lambda_1$. Then a solution of equation (5.1) from the class $C^q(V)$, $q \geq 1$, is the infinitesimal value which is characterized by the relation*

$$\Phi(z) = O(r^{\mu_{k_q+1}}), \quad r \to 0,$$

where $\mu_{k_q+1} = \sqrt{|\lambda|^2 + (k_q+1)^2}$ and k_q is defined by the inequality

$$\sqrt{q^2 - |\lambda|^2} - 1 < k_q \leq \sqrt{q^2 - |\lambda|^2}. \tag{5.9}$$

This theorem gives not only an asymptotic but also a numerical (estimating) formula. Namely, we have

Theorem 5.3 *Let $\lambda \in \Lambda_1, \Phi(z) \in C^q(V)$ and $|\Phi(z)| < M$ for all the points of the circle $\partial V = \{z : |z| = R\}$. Then for any point*

$$|\Phi(z)| < \frac{MR}{2(R-r)}\left(\frac{r}{\rho}\right)^{\mu_{k_q+1}}. \tag{5.10}$$

Theorem 5.4 *For any value $\lambda \in \Lambda_1$ equation (5.1) has no $C^\infty(G)$ solutions.*

Proof of Theorem 5.2 As $\lambda \in \Lambda_1$, then equality (5.8) is not possible. In that case if $\Phi(z) \in C^q(V)$, then in expression (4.5) it is necessary to set $A_0 = A_1 = \ldots = A_k = 0$, where the natural number k_q is defined by the inequality $\mu_{k_q} \leq q < \mu_{k_q+1}$. Resolving this with respect to k_q, we have (5.9). Theorem 5.2 is proved.

Proof of Theorem 5.3 From (4.6) we deduce the inequality

$$|A_k| < \frac{1}{2\mu_k}\left(1 + \frac{P_k}{|\lambda|}\right)\frac{M}{R^{\mu_k}}.$$

Further from (4.5) and by virtue of $\Phi(z) \in C^q(V)$ we have

$$|\Phi(z)| < \frac{1}{2}\left(\frac{r}{R}\right)^{\mu_{k_q+1}} \sum_{k_q=1}^{\infty} \left(\frac{r}{R}\right)^{\mu_k - \mu_{k_q+1}}.$$

Using the inequality $\mu_{k+s} - \mu_s \geq s$, we obtain the required estimate (5.10). Theorem 5.3 is proved.

Proof of Theorem 5.4 For a circular neighbourhood of the point $z = 0$ we have $\Phi(z) = 0$ from $\Phi(z) \in C^\infty$ and Theorem 5.3. But then $\Phi(z) \equiv 0$ in any neighbourhood of $z = 0$, since in view of the elliptic character of the problem outside the point $z = 0$ the function $\Phi(z)$ is analytic at all points z different from $z = 0$.

6 Analogy of Laurent series

We assume that equation (5.1) is considered in the circular ring $C = \{z : R_0 < |z| < R_1\}$. It is clear that the generalized integral Cauchy formula (2.5) will be valid in this case as well. For $z \in C$ formula (2.5) can be written in the form

$$\begin{aligned}\Phi(z) =& \frac{1}{2\pi i}\int_{\Gamma_1} \frac{\Omega_1(z,\zeta)}{\zeta}\Phi(\zeta)\,d\zeta - \frac{\Omega_2(z,\zeta)}{\bar{\zeta}}\overline{\Phi(\zeta)}\,\overline{d\zeta}\\ &-\frac{1}{2\pi i}\int_{\Gamma_0} \frac{\Omega_1(z,\zeta)}{\zeta}\Phi(\zeta)\,d\zeta - \frac{\Omega_2(z,\zeta)}{\bar{\zeta}}\overline{\Phi(\zeta)}\,\overline{d\zeta},\end{aligned}$$

where the circles $|z| = R_0$ and $|z| = R_1$ are denoted by Γ_0 and Γ_1, respectively. On Γ_0 and Γ_1 we have $\zeta = R_0 e^{i\theta_0}$ and $\zeta = R_1 e^{i\theta_1}$, so

$$\begin{aligned}\Phi(z) =& \frac{1}{2\pi}\int_0^{2\pi} \left\{\Omega_1(z, R_1 e^{i\theta_1})\Phi(R_1 e^{i\theta_1}) + \Omega_2(z, R_1 e^{i\theta_1})\overline{\Phi(R_1 e^{i\theta_1})}\right\} d\theta_1\\ &-\frac{1}{2\pi}\int_0^{2\pi} \left\{\Omega_1(z, R_0 e^{i\theta_0})\Phi(R_0 e^{i\theta_0}) + \Omega_2(z, R_0 e^{i\theta_0})\overline{\Phi(R_0 e^{i\theta_0})}\right\} d\theta_0.\end{aligned}$$

Introducing in place of Ω_1 and Ω_2 their expressions from (1.12), ch. 1 and realizing term–by–term integration, we get

$$\begin{aligned}\Phi(z) =& (a_0 + \frac{\lambda}{|\lambda|}\bar{a}_0)r^{|\lambda|} + \sum_{k=1}^{\infty}\left\{\frac{P_k a_k + \lambda \bar{a}_{-k}}{\mu_k}e^{ik\varphi} + \frac{P_{-k}a_{-k} + \lambda\bar{a}_k}{\mu_k}e^{-ik\varphi}\right\}r^{\mu_k}\\ &+(b_0 - \frac{\lambda}{|\lambda|}\bar{b}_0)r^{-|\lambda|} + \sum_{k=1}^{\infty}\left\{\frac{P_{-k}b_k - \lambda\bar{b}_{-k}}{\mu_k}e^{ik\varphi} + \frac{P_k b_{-k} - \lambda\bar{b}_k}{\mu_k}e^{-ik\varphi}\right\}r^{-\mu_k}\end{aligned} \tag{6.1}$$

where for values $k = 0, \pm 1, \ldots$

$$a_k = \frac{R_1^{-\mu_k}}{4\pi}\int_O^{2\pi}\Phi(R_1 e^{ik\theta_1})e^{-ik\theta_1}\,d\theta_1, \qquad b_k = \frac{R_0^{\mu_k}}{4\pi}\int_0^{2\pi}\Phi(R_0 e^{i\theta_0})e^{-ik\theta_0}\,d\theta_0.$$

The right part of formula (6.1) consists of two series sums: the first one contains terms only with positive degrees of r and the second one only with negative degrees of r. If we assume that $\Phi(z)$ on Γ_0 and Γ_1 is absolutely integrable with respect to θ_0 and θ_1, then these conditions will be sufficient for the absolute and uniform convergence of the first series with $|z| < R_1$ and the second series for $|z| > R_0$. As to the uniform convergence of the two series, this occurs in the ring $R_0 < |z| < R_1$.

If we call (4.1) the *power series analogy*, then we shall naturally name (6.1) the *Laurent series analogy* from the theory of analytic functions. By (6.1) the classification of possible singularities of the function $\Phi(z)$ at the point $z = 0$ is realized.

In fact, we assume that $\Phi(z)$ is a continuous solution of equation (5.1) in the domain $0 < |z| < R_1$. This domain is a degenerate circular ring with interior radius $R_O = O$. Series (6.1) play the part of the basic apparatus for investigation $\Phi(z)$ in a neighbourhood of $z = 0$. In a similar way to analytic function theory, the point $z = 0$ is named the μ_k *multiplicity pole* of the function $\Phi(z)$ if in a neighbourhood of $z = O$ the series expansion (6.1) does not contain terms with power order less than $-\mu_k$ and the coefficient for $r^{-\mu_k}$ is not equal to zero. If an infinite number of terms with a negative power of r is contained in (6.1) then the point $z = O$ is naturally called *the essential singular point* of the function $\Phi(z)$.

7 Generalized integral of Cauchy type

Let Γ be a simple smooth closed contour separating the plane E of a complex variable into two domains G^+ and G^- (contains the point $z = \infty$). Let us consider the integral

$$\Phi(z) = \frac{1}{2\pi i} \int_\Gamma \frac{\Omega_1(z,\zeta)}{\zeta} \nu(\zeta)\, d\zeta - \frac{\Omega_2(z,\zeta)}{\bar{\zeta}} \overline{\nu(\zeta)\, d\zeta}, \tag{7.1}$$

where Ω_1 and Ω_2 are the functions defined by formula (1.12) ch. 1, and $\nu(\zeta)$ is a function satisfying the Holder condition on Γ.

Integral (7.1) defines two independent functions $\Phi^+(z)$ and $\Phi^-(z)$ respectively in the domain G^+ and G^-. With regard to (2.9), ch.1, every one of them is a solution of equation (5.1); moreover $\Phi^+(z) \in C(G^+ + \Gamma) \cap C_{\bar{z}}(G^+ - 0)$, and $\Phi^-(z) \in C(G^- + \Gamma) \cap C_{\bar{z}}(G^-)$ and $\Phi^-(\infty) = O$. We shall call formula (7.1) *a generalized integral of Cauchy type*. In this section we establish some of its consequences.

7.1 Let us consider integral values of (7.1) on the integration contour ($\zeta_O \in \Gamma$)

$$\Phi(\zeta_O) = \frac{1}{2\pi i} \int_\Gamma \frac{\Omega_1(\zeta_O,\zeta)}{\zeta} \nu(\zeta)\, d\zeta - \frac{\Omega_2(\zeta_O,\zeta)}{\bar{\zeta}} \overline{\nu(\zeta) d\zeta}. \tag{7.2}$$

Starting from the properties of Ω_1, Ω_2 and the density $\nu(\zeta)$, and investigating them in a similar way to the integral of Cauchy type, see [23, 39], we get that (7.2) exists in the Cauchy general sense and defines the function $\Phi(\zeta_0)$, satisfying the Holder condition with the same exponent on Γ as $\nu(\zeta)$.

Now if in (7.1) we assume $z \in G^+$ (or G^-) and later pass over to the limit as $z \to \zeta_0 \in \Gamma$, then we get an analogy of the Sohockii–Plemelj formulas:

$$\Phi^+(\zeta_0) = \frac{1}{2}\nu(\zeta_0) + \Phi(\zeta_0), \qquad \Phi^-(\zeta_0) = \frac{1}{2}\nu(\zeta_0) + \Phi(\zeta_0), \tag{7.3}$$

where $\Phi(\zeta_0)$ is given by equality (7.2). Their obvious consequence are the relations:

$$\Phi^+(\zeta_0) - \Phi^-(\zeta_0) = \nu(\zeta_0), \qquad \Phi^+(\zeta_0) + \Phi^-(\zeta_0) = 2\Phi(\zeta_0).$$

7.2 Let us examine the problem of determining the piecewise regular function $\Phi(z)$ of (5.1) vanishing at $z = \infty$ by considering

$$\Phi^+(\zeta_0) - \Phi^-(\zeta_0) = \nu(\zeta_0) \qquad \text{on } \Gamma,$$

where $\nu(\zeta_0)$ is a given function satisfying the Holder condition. It is fairly clear that one of the solutions to the problem is defined by formula (7.1). We check that it is unique.

We assume that the problem has another solution. Let $\Psi(z)$ be the notation of their difference. It is obvious that $\Psi(z)$ is a solution of (5.1) and satisfies the equality $\Psi^+(\zeta_0) = \Psi^-(\zeta_0)$ on Γ. If we assign the corresponding values on Γ to the function $\Psi(z)$, then $\Psi(z)$ will belong to the class $C(E) \cap C_{\bar{z}}(E - O)$. Now by virtue of Theorem 4.2, $\Psi(z) \equiv O$ on the whole plane, and both the solutions coincide.

7.3 Let us study the following problem: *what kind of properties must a continuous function $\nu(\zeta), \zeta \in \Gamma$, satisfy to extend into the domain G^+ (or G^-) by the function which is a solution of equation* (5.1).

Theorem 7.1 *The equality*

$$\frac{1}{2\pi i}\int_\Gamma \frac{\Omega_1(z,\zeta)}{\zeta}\nu(\zeta)\,d\zeta - \frac{\Omega_2(z,\zeta)}{\bar{\zeta}}\overline{\nu(\zeta)\,d\zeta} = 0, \quad z \in G^-,$$

guarantees the continuous extendability of the function $\nu(\zeta), \zeta \in \Gamma$, into a function $\Phi^+(z), z \in G^+$, of equation (5.1) *from the class $C(G^+ + \Gamma) \cap C_{\bar{z}}(G^+ - O)$.*

Theorem 7.2 *The equality*

$$\frac{1}{2\pi i}\int_\Gamma \frac{\Omega_1(z,\zeta)}{\zeta}\nu(\zeta)\,d\zeta - \frac{\Omega_2(z,\zeta)}{\bar{\zeta}}\overline{\nu(\zeta)\,d\zeta} = 0, \quad z \in G^+,$$

guarantees the continuous extendability of the function $\nu(\zeta)$, $\zeta \in \Gamma$, *into a function* $\Phi^-(z), z \in G^-$, *of equation* (5.1) *from the class* $C(G^- + \Gamma) \cap C_{\bar{z}}(G^-)$.

The proof of these theorems is based on the generalized Cauchy integral properties and is not difficult.

7.4 We consider the integral equation

$$\frac{1}{\pi \mathrm{i}} \int_\Gamma \frac{\Omega_1(\zeta_O, \zeta)}{\zeta} \nu(\zeta)\, \mathrm{d}\zeta - \frac{\Omega_2(\zeta_O, \zeta)}{\bar{\zeta}} \overline{\nu(\zeta)\, \mathrm{d}\zeta} = \mu(\zeta_O), \tag{7.4}$$

where Γ is a smooth closed contour; ζ_O is an arbitrary point on Γ; Ω_1 and Ω_2 are defined by (1.12) ch. 1; $\nu(\zeta)$ is an unknown function and $\mu(\zeta)$ is a given function from the Holder class on Γ.

Theorem 7.3 *Equation* (7.4) *has a unique solution defined by the equality*

$$\nu(\zeta_O) = \frac{1}{\pi \mathrm{i}} \int_\Gamma \frac{\Omega_1(\zeta_O, \zeta)}{\zeta} \mu(\zeta)\, \mathrm{d}\zeta - \frac{\Omega_2(\zeta_O, \zeta)}{\bar{\zeta}} \overline{\mu(\zeta)\, \mathrm{d}\zeta}.$$

The proof of this theorem duplicates the scheme of resolving Cauchy–type integrals in the case of closed contours, see [39], §32.

8 Cases of a one–to–one correspondence between the sets $\{\Phi(z)\}$ and $\{\nu(\zeta)\}$.

Now we turn to the Cauchy–type generalized integral (7.1). If $\nu(\zeta)$ is given, $\Phi(z)$ can be uniquely calculated. The converse assertion, generally speaking, is not valid.

In this section we formulate cases of a one–to–one correspondence between the above sets, assuming that Γ is a circle $|z| < R$, and $\nu(\zeta)$ is expanded in an absolutely and uniformly convergent Fourier series with respect to $\theta = \arg \zeta$.

Introducing $\zeta = \mathrm{Re}^{\mathrm{i}\theta}$ into (7.1) and using the notation $\nu(\theta) = \nu(\mathrm{Re}^{\mathrm{i}\theta})$, we get

$$\Phi(z) = \frac{1}{2\pi} \int_O^{2\pi} \left\{ \Omega_1(z, \mathrm{Re}^{\mathrm{i}\theta}) \nu(\theta) + \Omega_2(z, \mathrm{Re}^{\mathrm{i}\theta}) \overline{\nu(\theta)} \right\} \mathrm{d}\theta.$$

Let $z \in G^+$. Applying formulas (1.12), ch. 1 and realizing term–by– term integration we have

$$\Phi^+(z) =(\nu_O + \frac{\lambda}{|\lambda|}\bar{\nu}_O)\Big(\frac{r}{R}\Big)^{|\lambda|} + \sum_{k=1}^{\infty}\Big\{\frac{P_k\nu_k + \lambda\bar{\nu}_{-k}}{\mu_k}e^{ik\varphi} + \frac{P_{-k}\nu_{-k} + \lambda\bar{\nu}_k}{\mu_k}e^{-ik\varphi}\Big\}\Big(\frac{r}{R}\Big)^{\mu_k}, \tag{8.1}$$

where $\nu_k = \dfrac{1}{4\pi}\displaystyle\int_O^{2\pi} \nu(\theta)e^{-ik\theta}\,d\theta, \quad k = 0, \pm 1, \ldots,$ series (8.1) converging absolutely and uniformly for $|z| \leq R$.

The correspondence between the sets of functions $\Phi^+(z)$ and $\nu(\zeta)$ will obviously be one–to–one, if the equality $\nu(\zeta) = O$ follows from the condition $\Phi^+(z) \equiv O$ and vice versa.

Thus, let $\Phi^+(z) = O$ be for all $z \in G^+$. Then from (8.1) we get

$$\nu_O + \frac{\lambda}{|\lambda|}\bar{\nu}_O = 0, \tag{8.2}$$

$$P_k\nu_k + \lambda\bar{\nu}_{-k} = 0, \quad k = 1, 2, \ldots . \tag{8.3}$$

We did not write out here the relation $P_{-k}\nu_{-k} + \lambda\bar{\nu}_k = O$, because it follows from (8.3) as was shown in subsection 4.4. Without dwelling on studying the arbitrariness in the determination of the Fourier coefficients $\nu_O, \nu_k, \nu_{-k}, k = 1, 2, \ldots$ we pass at once to the case, when $\nu(\theta)$ is a real function. Hence it is obvious that $\nu_O = \bar{\nu}_O, \nu_k = \bar{\nu}_k$, and equations (8.2) and (8.3) take the forms

$$\nu_O(1 + \frac{\lambda}{|\lambda|}) = O \qquad \nu_k(\lambda + P_k) = 0, \qquad k = 1, 2, \ldots .$$

Thus it follows that $\nu_O = O$ (except for the case when λ is a negative number) and $\nu_k = O, k = 1, 2, \ldots$.

If $\nu(\theta) = i\nu^*(\theta)$, where $\nu^*(\theta)$ is real, then from (8.2), (8.3) for the Fourier coefficients $\nu_O^*, \nu_k^* = \overline{\nu_{-k}^*}, k = 1, 2, \ldots$, of the function $\nu^*(\theta)$ we obtain the equations

$$\nu_O^*(1 - \frac{\lambda}{|\lambda|}) = O, \qquad \nu_k^*(P_k - \lambda) = O, \quad k = 1, 2, \ldots .$$

But this means that $\nu_O^* = O$ (the case of a positive λ is eliminated) and $\nu_k^* = O, k = 1, 2, \ldots$.

We summarize the above.

Theorem 8.1 *Let $\nu(\theta)$ be a real function. Then $\nu(\theta) = O$ follows from the condition $\Phi^+(z) \equiv O, |z| < R$, in such cases when $\lambda > O$* or Im $\lambda \neq O$; *in case $\lambda < O$ one has $\nu(\theta) = \nu_O$* where ν_O *is an arbitrary constant.*

Let $\nu(\theta) = \mathrm{i}\nu^*(\theta)$ be an imaginary function. Then $\nu(\theta) = O$ follows from the condition $\Phi^+(z) = O, |z| < R$, in such cases when $\lambda < O$ or $\mathrm{Im}\lambda \neq O$; in case $\lambda > O$ one has $\nu(\theta) = \mathrm{i}\nu^*_O$, *where ν^*_O is an arbitrary constant.*

Thus, Theorem 8.1 shows the cases when there exists a one–to–one correspondence between the sets of the functions $\Phi^+(z), z \in G^+$, and $\nu(z), z \in \Gamma$.

Now we write an expression for $\Phi^+(z)$ when $\nu(\theta)$ is a real function and when $\nu(\theta) = \mathrm{i}\nu^*(\theta)$ is a purely imaginary one. In the first case $\nu = \bar{\nu}_O, \nu_k = \bar{\nu}_{-k}$, therefore from (8.1) we get

$$\Phi^+(z) = (1+\frac{\lambda}{|\lambda|})\nu_O\left(\frac{r}{R}\right)^{|\lambda|} + \sum_{k=1}^{\infty}\left\{\frac{\lambda + P_k}{\mu_k}\nu_k \mathrm{e}^{\mathrm{i}k\varphi} + \frac{\lambda + P_{-k}}{\mu_k}\bar{\nu}_k \mathrm{e}^{-\mathrm{i}k\varphi}\right\}\left(\frac{r}{R}\right)^{\mu_k}. \quad (8.4)$$

In the second case $\nu_k = \mathrm{i}\nu^*_k(\nu_k = \bar{\nu}^*_{-k}), k = 1, 2, \dots$, therefore

$$\Phi^+(z) = \mathrm{i}(1 - \frac{\lambda}{|\lambda|})\nu^*_O\left(\frac{r}{R}\right)^{|\lambda|} + \mathrm{i}\sum_{k=1}^{\infty}\left\{\frac{P_k - \lambda}{\mu_k}\nu^*_k \mathrm{e}^{\mathrm{i}k\varphi} + \frac{P_{-k} - \lambda}{\mu_k}\bar{\nu}^*_k \mathrm{e}^{-\mathrm{i}k\varphi}\right\}\left(\frac{r}{R}\right)^{\mu_k}. \quad (8.5)$$

In a similar way we investigate cases of a one–to–one correspondence between the sets of the functions $\Phi^-(z), z \in G^-$, and $\nu(z), z \in \Gamma$. Without dwelling upon the proof we formulate the final result.

Theorem 8.2 *Let $\nu(\theta)$ be a real function. Then $\nu(\theta) = O$ follows from the condition $\Phi^-(z) = O, |z| \geq R$, in such cases when $\lambda < O$ or* Im $(\lambda) \neq O$; *in case $\lambda > O$ one has $\nu(\theta) = \nu_O$, where ν_O is an arbitrary constant.*

Let $\nu(\theta) = \mathrm{i}\nu^*(\theta)$ be an imaginary function. Then $\nu(\theta) = O$ follows from the condition $\Phi^-(z) = O, |z| \geq R$, when $\lambda > O$ or Im $\lambda \neq O$; *in case $\lambda < O$ one has $\nu(\theta) = \mathrm{i}\nu^*_O$, where ν^*_O is an arbitrary constant.*

9 Riemann–Hilbert problem for solutions of the model equation

9.1 We denote the disk $|z| < R$ by G and the circle $r = Re^{\mathrm{i}\varphi}$ by Γ. Let us consider the boundary value problem

$$\partial_{\bar{z}}\Phi - \frac{\lambda}{2\bar{z}}\bar{\Phi} = O, \quad z \in G, \quad (9.1)$$

$$\mathrm{Re}[z^{-m}\Phi] = h(z), \quad z \in \Gamma, \tag{9.2}$$

where $\Phi(z)$ is an unknown function from the class $C(G+\Gamma) \cap C_{\bar{z}}(G-O), n$ is an integer, and $h(z)$ is a given function. We assume that $h(\varphi) = h(Re^{ik\varphi})$ is expanded in an absolutely and uniformly convergent Fourier series

$$h(\varphi) = h_O + \frac{1}{2}\sum_{k=1}^{\infty} h_k e^{ik\varphi} + \bar{h}_k e^{-ik\varphi}, \tag{9.3}$$

where

$$h_O = \frac{1}{2\pi}\int_O^{2\pi} h(\varphi)\,d\varphi, \qquad h_k = \frac{1}{\pi}\int_O^{2\pi} h(\varphi)e^{-ik\varphi}\,d\varphi \qquad (k = 1, 2\ldots).$$

Investigation of the boundary value problem can be carried out by formulas (8.4), (8.5); in addition (8.4) is applied when $\lambda > O$ or $\mathrm{Im}\ \lambda \neq O$, and (8.5) is used when $\lambda < O$. It is then that the one–to–one correspondence between the sets $\Phi^+(z)$ and $\nu(\theta)$ exists according to Theorem 8.1. Introducing the above formulas into the boundary condition (9.2) and determining the unknown Fourier coefficients of the functions $\nu(\theta)$, we can address the whole solvability situation of the boundary value problem. This method, however, is not convenient since it involves an awkward computation stipulated by utilizing the two formulas (8.4) and (8.5). So the study of the problem will be based on the representation of the solutions to the model equation in the form

$$\Phi(z) = \mathcal{X} a_O \left(\frac{r}{R}\right)^{|\lambda|} \sum_{k=1}^{\infty}\left(\lambda a_k e^{ik\varphi} + P_{-k}\bar{a}_k e^{-ik\varphi}\right)\left(\frac{r}{R}\right)^{\mu_k}, \tag{9.4}$$

which has been obtained in §4, see (4.5), and combines formulas (8.4) and (8.5) together.

Further the cases $m = O, m > O$ and $m < O$ in condition (9.2) will be separately considered.

Theorem 9.1 $(m = O)$. *The problem* (9.1), (9.2) *for* $\lambda > O$ *or* $\mathrm{Im}\ \lambda \neq O$ *has the unique solution*

$$\Phi(z) = h_O \frac{\mathcal{X}}{\mathrm{Re}\mathcal{X}}\left(\frac{r}{R}\right)^{|\lambda|} + \sum_{k=1}^{\infty}\left(\frac{\lambda h_k}{\lambda + P_{-k}} e^{ik\varphi} + \frac{P_{-k}\bar{h}_k}{\bar{\lambda} + P_{-k}} e^{-ik\varphi}\right)\left(\frac{r}{R}\right)^{\mu_k}.$$

For $\lambda < O$ the homogeneous problem ($h = O$) has the solution $icr^{-\lambda}$, where c is an arbitrary constant, and to solve the inhomogeneous problem it is necessary and sufficient that the condition $h_O = O$ be satisfied. The general solution of the problem is given by the formula

$$\Phi(z) = icr^{-\lambda} + H(z),$$

where

$$H(z) = \sum_{k=1}^{\infty} \left(\frac{\lambda h_k}{\lambda + P_{-k}} e^{ik\varphi} + \frac{P_{-k}\bar{h}_k}{\bar{\lambda} + P_{-k}} e^{-ik\varphi} \right) \left(\frac{r}{R} \right)^{\mu_k}.$$

Proof Substituting (9.3) and (9.4) into the boundary condition (9.2) and equating coefficients for equal degrees of $e^{ik\varphi}$, we arrive at a system of algebraic equations for determining the unknown coefficients a_O and a_k:

$$a_O \mathrm{Re}\mathcal{X} = h_O, \qquad (\lambda + P_{-k})a_k = h_k \qquad (k = 1, 2, \ldots). \tag{9.5}$$

Hence it is obvious that for any λ the coefficients a_k are uniquely defined. The same assertion is true for a_O if $\lambda > O$ or Im $\lambda \neq O$ (then Re $\mathcal{X} \neq O$). In the case when $\lambda < O$ we have Re $\mathcal{X} = O$ (as $\mathcal{X} = i$), and a_O becomes arbitrary, and $h_O = O$.

This makes evident the assertion of Theorem 9.1. Introducing the coefficients a_O and a_k, calculated from (9.5), into (9.4), we get a series which is a solution of the boundary value problem (9.1), (9.2). The absolute and uniform convergence of this series follows from the absolute convergence of the Fourier series for $h(\varphi)$. Theorem 9.1 is proved.

9.2 The case $m > O$. Substituting (9.4) into (9.2) leads to an infinite system of algebraic equations for determining the constants a_O, a_k:

$$\mathrm{Re}(\lambda a_m) = h_O R^m, \tag{9.6}$$

$$\lambda a_{m+p} + \overline{\lambda a}_{m-p} = h_p R^m \quad (p = 1, \ldots, m-1), \tag{9.7}$$

$$\lambda a_{2m} + \bar{\mathcal{X}} a_O = h_m R^m, \tag{9.8}$$

$$\lambda a_{2m+k} + P_{-k} a_k = h_{m+k} R^m \quad (k = 1, 2, \ldots). \tag{9.9}$$

First we shall study the homogeneous problem (9.1), (9.2). To this end we assume $h(z) = O$ in the condition (9.2). This problem corresponds to the infinite system of homogeneous algebraic equations which one gets from (9.6) – (9.9) provided that $h_k = O, k = O, 1, \ldots$, and which it is convenient to denote $(9.6^0) - (9.9^0)$.

From (9.9^0) one determines $a_m = \mathrm{i}\bar{\lambda}c_m$, where c_m is an arbitrary real constant. Knowing the value of a_m, one can determine all the constants $a_{(2k+1)m}, k = 1, 2, \ldots$, by the sequence of formulas

$$\lambda a_{(2k+1)m} + P_{-(2k-1)m} a_{(2k-1)m} = O, \tag{9.10}$$

which follow from (9.9^0). From (9.10) we get that

$$a_{(2k+1)m} = \frac{1}{(-\lambda)^k} \cdot \prod_{\delta=1}^{k} P_{-(2\delta-1)m} \cdot \mathrm{i}\bar{\lambda}c_m \quad (k = 1, 2, \ldots),$$

where $\prod$ is the product symbol.

The arbitrary constant c_m generates a non-trivial solution of the homogeneous boundary problem

$$\Phi_m(z) = \sum_{k=0}^{\infty} \left[\lambda a_{(2k+1)m} e^{\mathrm{i}(2k+1)m\varphi} + P_{-(2k+1)m} \bar{a}_{(2k+1)m} e^{-\mathrm{i}(2k+1)m\varphi} \right] \left(\frac{r}{R} \right)^{\mu_{(2k+1)m}}. \tag{9.11}$$

Here in place of the coefficients $a_{(2k+1)m}$ one uses their expressions in terms of c_m. It is not hard to prove the absolute and uniform convergence of the series (9.11) in the closed disk $|z| \leq R$.

Now we turn to equation (9.7^O). If one believes $a_{m-p} = c_{m-p} (p = 1, \ldots, m-1)$, where c_{m-p} is an arbitrary complex constant, then it will follow from (9.7^O) that $a_{m+p} = -(\bar{\lambda}/\lambda)\bar{c}_{m-p}$ and from the recurrence relations

$$\lambda a_{(2k+1)m-p} + P_{-[(2k+1)m-p]} a_{(2k-1)m-p} = 0,$$

$$\lambda a_{(2k+1)m+p} + P_{-[(2k-1)m+p]} a_{(2k-1)m+p} = 0, \quad k = 1, 2, \ldots,$$

which issue from (9.9^O), that one can compute all co efficients of the form $a_{(2k+1)m\pm p}, k = 1, 2, \ldots$:

$$a_{(2k+1)m-p} = \frac{1}{(-\lambda)^k} \prod_{\delta=1}^{k} P_{-[(2\delta-1)m-p]} c_{m-p}, \tag{9.12$^-$}$$

$$a_{(2k+1)m+p} = \frac{1}{(-\lambda)^k} \prod_{\delta=1}^{k} P_{-[(2\delta-1)m+p]} \left(-\frac{\bar{\lambda}}{\lambda} \right) c_{m-p}. \tag{9.12$^+$}$$

The arbitrary constants $c_{m-p}(p = 1, \ldots, m-1)$ generate non–trivial solutions of the homogeneous problem (9.1), (9.2):

$$\begin{aligned}\Phi_{m-p}(z) = \sum_{k=0}^{\infty} \Big\{ & \lambda a_{(2k+1)m-p} e^{i[(2k+1)m-p]\varphi} \\ & + P_{-[(2k+1)m-p]} \bar{a}_{(2k+1)m-p} e^{-i[(2k+1)m-p]\varphi} \Big\} \left(\frac{r}{R}\right)^{\mu_{(2k+1)m-p}} \\ & + \Big\{ \lambda a_{(2k+1)m+p} e^{i[(2k+1)m+p]\varphi} \\ & + P_{-[(2k+1)m+p]} \bar{a}_{(2k+1)m+p} e^{-i[(2k+1)m+p]\varphi} \Big\} \left(\frac{r}{R}\right)^{\mu_{(2k+1)m+p}}\end{aligned} \tag{9.13}$$

where $a_{(2k+1)m+p}$ are expressed through c_{m-p} by formulas (9.12). The series (9.13) converges absolutely and uniformly for $|z| \le R$.

Finally, we consider equation (9.8^O). Assuming $a_O = c_O$(c_O is an arbitrary real constant), we have $a_{2m} = -(1/\lambda)\mathcal{X} c_O$, and using the recurrence relation

$$\lambda a_{2(k+1)m} + P_{-2mk} a_{2mk} = O \quad (k = 1, 2, \ldots),$$

which follows from (9.9^O), we determine uniquely all the coefficients of the type $a_{2m(k+1)}$:

$$a_{2m(k+1)} = \frac{1}{(-\lambda)^k} \prod_{\delta=1}^{k} P_{-2m\delta} \bar{\mathcal{X}} c_O. \tag{9.14}$$

There is still another non–trivial solution of problem (9.1), (9.2) connected with the constant c_O, viz.

$$\Phi_O(z) = \mathcal{X} c_O \left(\frac{r}{R}\right)^{|\lambda|} + \sum_{k=1}^{\infty} \left[\lambda a_{2mk} e^{2imk\varphi} + P_{-2mk} \bar{a}_{2mk} e^{-2imk\varphi} \right] \left(\frac{r}{R}\right)^{\mu_{2mk}}, \tag{9.15}$$

where in place of a_{2mk} one uses its expression (9.14) in terms of c_O. It is easy to verify that the series (9.15) converges absolutely and uniformly for $|z| \le R$.

Thus, the general solution $\Phi^*(z)$ of the homogeneous boundary value problem (9.1), (9.2) has the form

$$\Phi^*(z) = \Phi_O(z) + \sum_{p=1}^{m-1} \Phi_{m-p}(z) + \Phi_m(z). \tag{9.16}$$

If we denote

$$\tilde{\Phi}_O(z) = [\Phi_O(z)]_{c_O=1}; \qquad \tilde{\Phi}_{m-p}(z) = [\Phi_{m-p}(z)]_{c_{m-p}=1};$$

$$\tilde{\tilde{\Phi}}_{m-p}(z) = [\Phi_{m-p}(z)]_{c_{m-p}=i}; \qquad \tilde{\Phi}_m(z) = [\Phi_m(z)]_{c_m=1},$$

then the general solution can be written as follows:

$$\Phi^*(z) = c_O\tilde{\Phi}_O(z) + \sum_{p=1}^{m-1} c_{m-p}^{(1)}\tilde{\Phi}_{m-p}(z) + c_{m-p}^{(2)}\tilde{\tilde{\Phi}}_{m-p}(z) + c_m\tilde{\Phi}_m(z), \tag{9.17}$$

where

$$c_{m-p}^{(1)} = \operatorname{Re} c_{m-p} \text{ and } c_{m-p}^{(2)} = \operatorname{Im} c_{m-p}.$$

So, the homogeneous problem (9.1), (9.2) has $2m$ linearly independent solutions. The totality of all solutions is given by either of the formulas (9.16) or (9.17).

Now we turn to the inhomogeneous boundary problem (9.1), (9.2). One of its particular solutions will be

$$H(z) = \sum_{p=0}^{\infty} H_p(z), \tag{9.18}$$

where

$$H_p(z) = \sum_{k=0}^{\infty}\Bigg[\lambda b_{(2k+1)m+p}^{(p)} e^{i[(2k+1)m+p]\varphi} + P_{-[(2k+1)m+p]} b_{(2k+1)m+p}^{(p)} e^{-i[(2k+1)m+p]\varphi}\Bigg]\left(\frac{r}{R}\right)^{\mu_{(2k+1)m+p}},$$

while for any values $p = 0, 1, \ldots$

$$b_{m+p}^{(p)} = \frac{1}{\lambda} h_p R^m, \qquad b_{(2k+1)m+p}^{(p)} = \frac{1}{(-\lambda)^k}\prod_{\delta=1}^{k} P_{-[(2\delta-1)m+p]}\frac{1}{\lambda}R^m h_p.$$

An immediate verification shows that for any values of the indices p and k separate summands of the series which define $H_p(z)$ are a solution of equation (9.1). Therefore $H_p(z), p = 0, 1, \ldots$, will also be a solution, as will $H(z)$.

An immediate verification also proves that

$$\operatorname{Re}\left[z^{-m}H_p\right] = \frac{1}{2}\left(h_p e^{ip\varphi} + \bar{h}_p e^{-ip\varphi}\right) \quad \text{on } |z| = R.$$

Further, taking (9.18) into account, we find that $H(z)$ satisfies (9.2).

We shall now prove the absolute and uniform convergence of the series (9.18). Its separate summands satisfy the inequalities

$$|H_p(z)| \leq R^m a_p |h_p|, \quad p = 0, 1, \dots,$$

where

$$a_p = 1 + 2\sum_{k=1}^{\infty} |\lambda|^{-k} \prod_{\delta=1}^{k} P_{-[(2\delta-1)m+p]}.$$

The important inequality $a_{p_2} < a_{p_1}$ for $p_2 > p_1 \geq O$ issues from the inequality $P_{-k_2} < P_{-k_1}$ for $k_2 > k_1 \geq O$ (we recall that $P_k = (|\lambda|^2 + k^2)^{1/2} - k$). This shows that the constants $a_p (p = O, 1, \dots)$ are bounded above: $a_p \leq a_O$. Now from (9.18) we get

$$|H(z)| \leq \sum_{p=O}^{\infty} |H_p(z)| \leq R^m \sum_{p=O}^{\infty} a_p |h_p| < a_O R^m \sum_{p=O}^{\infty} |h_p|.$$

Hence the uniform convergence of (9.18) follows. So we have

Theorem 9.2 *($m > O$ case). The homogeneous problem (9.1), (9.2) ($h = O$) has $2m$ linearly independent solutions and the inhomogeneous problem is always solvable. Its general solution can be given in the form*

$$\Phi(z) = \Phi^*(z) + H(z)$$

where $\Phi^*(z)$ is given by any of the formulas (9.16) or (9.17), and $H(z)$ by formula (9.18).

9.3 Now we consider the case $m < O$.

Theorem 9.3 *The homogeneous problem has only the zero solution and for the existence of a solution of the inhomogeneous problem it is necessary and sufficient that $2|m|$ real solvability conditions are satisfied:*

$$h_O + \mathrm{Re}\left(\sum_{\beta=1}^{\infty} \frac{(-1)^\beta}{\lambda^\beta} h_{2\beta|m|} \prod_{\delta=O}^{\beta-1} P_{-(2\delta+1)|m|} \right) = 0, \tag{9.19$_0$}$$

$$h_O + \sum_{\beta=1}^{\infty} \frac{(-1)^\beta}{\lambda^\beta} h_{2\beta|m|+p} \prod_{\delta=O}^{\beta-1} P_{-[(2\delta+1)|m|+p]} \tag{9.19$_p$}$$

$$+ \sum_{\beta=1}^{\infty} \frac{(-1)^\beta}{\bar{\lambda}^\beta} h_{2\beta|m|-p} \prod_{\delta=O}^{\beta-1} P_{-[(2\delta+1)|m|-p]} = O \quad (p = 1, \dots, |m| - 1)$$

$$\mathrm{Im}\frac{1}{\mathcal{X}}\left(h_{|m|}+\sum_{\beta=1}^{\infty}\frac{(-1)^{\beta}}{\lambda^{\beta}}h_{(2\beta+1)|m|}\prod_{\delta=O}^{\beta-1}P_{-(2\delta+2)|m|}\right)=0. \tag{9.19$_\mathrm{m}$}$$

Under these conditions the problem has the following solution

$$\begin{aligned}\Phi(z)=&\sum_{p=0}^{|m|-1}\sum_{k=O}^{\infty}\Big(\lambda a_{(2k+1)|m|-p}e^{i[(2k+1)|m|-p]\varphi}\\&+P_{-[(2k+1)|m|-p]}\bar{a}_{(2k+1)|m|-p}e^{-i[(2k+1)|m|-p]\varphi}\Big)\left(\frac{r}{R}\right)^{\mu(2(k+1)|m|-p}\\&+\sum_{p=1}^{|m|}\sum_{k=O}^{\infty}\Big(\lambda a_{(2k+1)|m|+p}e^{i[(2k+1)|m|+p]\varphi}\\&+P_{-[(2k+1)|m|+p]}\bar{a}_{(2k+1)|m|+p}e^{-i[(2k+1)|m|+p]\varphi}\Big)\left(\frac{r}{R}\right)^{\mu(2k+1)|m|+p}\\&+\mathcal{X}a_O\left(\frac{r}{R}\right)^{|\lambda|},\end{aligned} \tag{9.20}$$

where

$$a_O=R^m\mathrm{Re}\frac{1}{\mathcal{X}}\left(h_{|m|}+\sum_{\gamma=1}^{\infty}\frac{(-1)^{\gamma}}{\lambda\gamma}h_{(2\gamma+1)|m|}\prod_{\delta=1}^{\gamma}P_{-2\delta|m|}\right),$$

$$\begin{aligned}a_{(2k+1)|m|\pm p}=&\frac{R^m}{\lambda}\Bigg(h_{2(k+1)|m|\pm p}+\sum_{\gamma=1}^{\infty}\frac{(-1)^{\gamma}}{\lambda\gamma}h_{2(\gamma+k+1)|m|\pm p}\\&\times\prod_{\delta=1}^{\gamma}P_{-[(2\delta+2k+1)|m|\pm p]}\Bigg).\end{aligned}$$

Proof Substituting (9.4) into (9.2) and equating coefficients of the same powers of $e^{i\varphi}$, we get

$$P_{-|m|}\mathrm{Re}a_{|m|}=h_O R^m, \tag{9.21}$$

$$P_{-(|m|+p)}a_{|m|+p}+P_{-(|m|-p)}\bar{a}_{|m|-p}=h_pR^m\quad(p=1,\ldots,|m|-1), \tag{9.22}$$

$$P_{-2|m|}a_{2|m|}+\mathcal{X}a_O=h_{|m|}R^m, \tag{9.23}$$

$$P_{-(2|m|+k)}a_{2|m|+k}+\lambda a_k=h_{|m|+k}R^m\quad(k=1,2,\ldots). \tag{9.24}$$

From (9.21) we define

$$a_{|m|}=\frac{R^m}{P_{-|m|}}(ic_{|m|}+h_O),$$

where $c_{|m|}$ is an arbitrary real constant and then from (9.24) which is convenient to write in the form

$$P_{-(2k+1)|m|}a_{(2k+1)|m|}+\lambda a_{(2k-1)|m|}=R^m h_{2k|m|}\quad (k=1,2,\ldots),$$

we compute

$$a_{(2k+1)|m|}=\frac{(-\lambda)^k R^m}{\prod_{\delta=0}^{k}P_{-(2\delta+1)|m|}}\left(\mathrm{i}c_{|m|}+\sum_{\beta=0}^{k}\frac{h_{2\beta|m|}}{(-\lambda)^\beta}\prod_{\delta=0}^{\beta-1}P_{-(2\delta+1)|m|}\right).\qquad (9.25)$$

Now we consider equation (9.22). Assuming

$$a_{|m|-p}=\frac{R^m}{P_{-(|m|-p)}}c_{|m|-p}\quad (p=1,\ldots,|m|-1),$$

where $c_{|m|-p}$ is an arbitrary complex constant, we obtain

$$a_{|m|+p}=\frac{R^m}{P_{-(|m|+p)}}(-c_{|m|-p}+h_p)\quad (p=1,\ldots,|m|-1).$$

From (9.24) we extract the recurrence formula

$$P_{-[(2k+1)|m|\pm p]}a_{(2k+1)|m|\pm p}+\lambda a_{(2k-1)|m|\pm p}=R^m h_{2k|m|\pm p},$$

which helps to define the coefficients $a_{(2k+1)|m|\pm p}, k=1,2,\ldots,$ through the known constants $a_{|m|-p}$ and $a_{|m|+p}$

$$a_{(2k+1)|m|-p}=\frac{(-\lambda)^k R^m}{\prod_{\delta=0}^{k}P_{-[(2\delta+1)|m|-p]}}\left(c_{|m|-p}+\sum_{\beta=1}^{k}\frac{h_{2\beta|m|-p}}{(-\lambda)^\beta}\right.$$
$$\left.\times\prod_{\delta=0}^{\beta-1}P_{-[(2\delta+1)|m|-p]}\right)\qquad (9.26^-)$$

$$a_{(2k+1)|m|+p}=\frac{(-\lambda)^k R^m}{\prod_{\delta=0}^{k}P_{-[(2\delta+1)|m|+p]}}\left(h_p-\bar{c}_{|m|-p}+\sum_{\beta=1}^{k}\frac{h_{2\beta|m|+p}}{(-\lambda)^\beta}\right.$$
$$\left.\times\prod_{\delta=0}^{\beta-1}P_{-[(2\delta+1)|m|+p]}\right).\qquad (9.26^+)$$

Finally, fixing the coefficient a_O in (9.23) by the equality $a_O = c_O R^m$, where c_O is an arbitrary real constant, we have

$$a_{2|m|} = \frac{R^m}{P_{-2|m|}}(h_m - \mathcal{X} c_O).$$

Then we deduce from (9.24) the recurrence formula

$$P_{-(2k+1)|m|} a_{(2k+1)|m|} + \lambda a_{2k|m|} = R^m h_{(2k+1)|m|},$$

and compute with it the coefficients $a_{(2k+1)|m|}, k = 1, 2, \dots$:

$$a_{(2k+1)|m|} = \frac{(-\lambda)^k R^m}{\prod_{\delta=O}^{k} P_{-(2\delta+1)|m|}} \left(h_{|m|} - \mathcal{X} c_O + \sum_{\beta=1}^{k} \frac{h_{(2\beta+1)|m|}}{(-\lambda)^\beta} \prod_{\delta=O}^{\beta-1} P_{-(2\delta+1)|m|} \right). \tag{9.27}$$

Thus, we could express all the coefficients $a_k (k = O, 1, \dots)$ in formula (9.4) in terms of the arbitrary real constants $c_O, c_{|m|}$ and $|m| - 1$ arbitrary complex constants $c_1, \dots, c_{|m|-1}$, and the Fourier coefficients $h_O, h_k (k = 1, 2, \dots)$ of the function $h(z), |z| = R$, as well.

The next stage of the proof of Theorem 9.3 is the study of the series convergence. First we examine the series (9.4) in which only the coefficients of the type $a_{(2k+1)|m|}, k = 1, 2, \dots$, are different from zero:

$$\Phi_{|m|}(z) = \sum_{k=O}^{\infty} A_{q(k)} \left(\frac{r}{R} \right)^{\mu_{q(k)}}, \tag{9.28}$$

where $q(k) = (2k+1)|m|$ and

$$A_{q(k)} = \lambda a_{q(k)} e^{iq(k)\varphi} + P_{-q(k)} \bar{a}_{q(k)} e^{-iq(k)\varphi}. \tag{9.29}$$

This series formally satisfies equation (9.1) and boundary condition (9.2) in which all the Fourier coefficients of $h(\varphi)$ except $h_{2k|m|}, k = 0, 1, \dots$, are equal to zero.

Since we are interested only in the convergent series (9.28) then we must first check the necessary condition for convergence:

$$\lim_{k \to \infty} A_{q(k)} \left(\frac{r}{R} \right)^{\mu_{q(k)}} = O, \quad |z| \le R. \tag{9.30}$$

Using (9.29), (9.25), the notation $P_{-s} = \sqrt{|\lambda|^2 + s^2} - s$ which has been applied, and the obvious relation

$$\frac{1}{|\lambda|} P_{-s} = \frac{|\lambda|}{\sqrt{|\lambda|^2 + s^2} + s}, \tag{9.31}$$

we find that (9.30) can be realized only when $\lim\limits_{k\to\infty} a_{(2k+1)|m|} = 0$, i.e.

$$ic_{|m|} + h_O + \sum_{\beta=1}^{\infty} \frac{h_{2\beta|m|}}{(-\lambda)^\beta} \prod_{\delta=O}^{\beta-1} P_{-(2\delta+1)|m|} = 0. \tag{9.32}$$

Separating the real and imaginary parts, we get two real conditions. The first of them is already written under (9.19_0); the second has the form:

$$c_{|m|} + \mathrm{Im}\left(\sum_{\beta=1}^{\infty} \frac{h_{2\beta|m|}}{(-\lambda)^\beta} \prod_{\delta=O}^{\beta-1} P_{-(2\delta+1)|m|} \right) = 0.$$

This relation is intended for determining the arbitrary constant $c_{|m|}$. In a particular case when we mean the homogeneous boundary problem (9.1), (9.2) ($h(\varphi) = O$), we obviously have $c_{|m|} = O$, and from (9.25) we get $a_{(2k+1)|m|} = O, k = O, 1, \ldots$. This means that the constant $c_{|m|}$ does not generate a non–trivial solution of the boundary problem.

Now we assume that condition (9.32) is fulfilled. With its help we transform (9.25) to the form

$$a_{(2k+1)|m|} = \frac{R^m}{\lambda} \left(h_{2(k+1)|m|} + \sum_{\gamma=1}^{\infty} \frac{h_{2(\gamma+k+1)|m|}}{(-\lambda)^\gamma} \prod_{\delta=1}^{\gamma} P_{-(2\delta+2k+1)|m|} \right).$$

Using this equality in formula (9.28) and omitting the simple intermediate calculation, we get

$$|\Phi_{|m|}(z)| < \sum_{p=1}^{\infty} a_{2p|m|} |h_{2p|m|}|, \tag{9.33}$$

where $a_{2|m|} = 1$ and

$$a_{2(p+1)|m|} = 1 + \frac{P_{-2(p+1)|m|}}{|\lambda|} a_{2p|m|} \quad (p = 1, 2, \ldots).$$

Hence, on the basis of (9.31), it is easy to state that $a_{2p|m|} (p = 1, 2 \ldots)$ is bounded above by one and the same number. But then from (9.33) and the absolute convergence of the Fourier series for $h(\varphi)$ the uniform convergence of series (9.28) and continuity of its sum $\Phi_{|m|}(z)$ follow for $|z| \le R$.

Below we analyse the series $\Phi_{|m|-p}(z)$ and $\Phi_O(z)$, which are generated by (9.4) with the coefficients $a_{(2k+1)|m|\pm p}$ and $a_{(2k+1)|m|}$ different from zero respectively. Reasoning in an analogous way we establish that the necessary condition for their convergence leads to the demands

$$\lim_{k\to\infty} a_{(2k+1)|m|\pm p} = \lim_{k\to\infty} a_{2(k+1)|m|} = O.$$

According to (9.26$^\pm$) and (9.27) this is equivalent to the relations:

$$c_{|m|-p} + \sum_{\beta=1}^{k} \frac{h_{2\beta|m|-p}}{(-\lambda)^\beta} \prod_{\delta=O}^{\beta-1} P_{-[(2\delta+1)|m|-p]} = 0, \tag{9.34$^-$}$$

$$h_p - \bar{c}_{|m|-p} + \sum_{\beta=1}^{k} \frac{h_{2\beta|m|+p}}{(-\lambda)^\beta} \prod_{\delta=0}^{\beta-1} P_{-[(2\delta+1)|m|+p]} = 0, \tag{9.34$^+$}$$

$$h_{|m|} - \mathcal{X} c_O + \sum_{\beta=1}^{k} \frac{h_{2\beta+1)|m|}}{(-\lambda)^\beta} \prod_{\delta=0}^{\beta-1} P_{-2(\delta+1)|m|} = 0. \tag{9.35}$$

The equality (9.34$^-$) serves for the determination of the arbitrary constant $c_{|m|-p}$. Ignoring it in (9.34$^+$) we come to the solvability condition (9.19$_p$).

As for (9.35) we divide it first by $\mathcal{X}$ and then separate its real and imaginary parts. One establishes that the first of them is destined for determining the unknown real constant c_O, and the second gives the necessary solvability written by formula (9.19$_\mathrm{m}$).

Relations (9.34$^\pm$), (9.35) indicate that the constants $c_O, c_{|m|\pm p}$ do not generate non–trivial solutions of the homogeneous boundary problem (they are equal to zero for $h(\varphi) = 0$). These relations are not only necessary but also sufficient to solve the boundary problem. To prove this, formulas (9.25), (9.26$^\pm$) should be transformed beforehand by (9.34$^\pm$), (9.35). The final expression for the coefficients has been written out in Theorem 9.3 after (9.20). The uniform convergence of the series $\Phi_O(z)$ and $\Phi_{|m|-p}(z)$ is checked by a similar scheme to $\Phi_{|m|}(z)$.

It remains to state that the sum

$$\Phi_O(z) + \sum_{p=1}^{|m|-1} \Phi_{|m|-p}(z) + \Phi_{|m|}(z)$$

gives the unique continuous solution of boundary problem (9.1), (9.2) in the domain $|z| \leq R$, its explicit form being expressed in Theorem 9.3 under (9.20). Theorem 9.3 is proved.

9.4 This point is a supplement to the previous one. The solvability conditions (9.19) are expressed in the form

$$\int_0^{2\pi} \mathrm{e}^{\mathrm{i}(m+1)\varphi} U_k(R\mathrm{e}^{\mathrm{i}\varphi}) h(\varphi)\, \mathrm{d}\varphi = 0, \quad k = 0, 1, \dots, |m|,$$

$$\int_0^{2\pi} \mathrm{e}^{\mathrm{i}(m+1)\varphi} U_k^*(R\mathrm{e}^{\mathrm{i}\varphi}) h(\varphi)\, \mathrm{d}\varphi = 0, \quad k = 1, \dots, |m| - 1. \tag{9.36}$$

where the functions $U_O(z), U_{|m|}(z), U_k(z), U_k^*(z), k = 1, \dots, |m| - 1$, make up the complete system of linearly independent solutions (as a combination with real coefficients) of the homogeneous boundary problem

$$\partial_{\bar{z}} U - \frac{\bar{\lambda}}{2z} \bar{U} = 0, \quad |z| < R, \tag{9.37}$$

$$\mathrm{Re}\left[\mathrm{i}\mathrm{e}^{\mathrm{i}(m+1)\varphi} U\right] = 0 \quad \text{on} \quad |z| = R, \tag{9.38}$$

the solution being found from the class of functions continuous and continuously differentiable outside the point $z = 0$ and allowing a first-order singularity at $z = O$.

To examine this assertion it is necessary to construct a system of linearly independent solutions for problem (9.37), (9.38). In this connection we should turn to the expressions Ω_1^* and Ω_2^* from (1.2), write out, on their basis, the generalized integral of Cauchy type (7.1) and then the solution representation of equation (9.37) in the form of a power series analogy. Finally we get

$$U(z) = \sum_{k=-1}^{\infty} \left\{ -\bar{\lambda} b_k \mathrm{e}^{\mathrm{i}k\varphi} + Q_k \bar{b}_k \mathrm{e}^{-\mathrm{i}(k+2)\varphi} \right\} r^{\nu_k}, \tag{9.39}$$

where $\nu_k = \sqrt{(k+1)^2 + |\lambda|^2} - 1$, $Q_k = \nu_k - k, b_k$ are arbitrary complex constants. Generally speaking, this expression has been written out for λ values satisfying the inequality $1 \leq |\lambda| \leq \infty$. In the case $0 < |\lambda| < 1$ the summation in (9.39) starts from $k = 0$, i.e. $b_{-1} = 0$.

Further study of problem (9.37), (9.38) is similar to the previous subsections. To avoid redundant awkwardness we do not write out the explicit form of its linearly independent solutions. These are characterized by two properties. The first one indicates the behaviour of the solution as $z \to O$:

$$U_0(z) = O(|z|^{\mu_0-1}), \qquad U_{|m|}(z) = O(|z|^{\mu_{|m|}-1}),$$
$$U_k(z), U_k^*(z) = O(|z|)^{m_k-1}), \qquad \mu_k^2 = |\lambda|^2 + k^2.$$

The description of the second property is conveniently realized through the additional functions $\mathbf{u}_k(\varphi)$ and $\nu_k(\varphi), k = 0, 1 \ldots, |m|$, which are defined in the following way:

$$\mathbf{u}_O(\varphi) = U_O(\mathrm{Re}^{\mathrm{i}\varphi}), \qquad u_{|m|}\varphi = U_{|m|}(\mathrm{Re}^{\mathrm{i}\varphi}),$$

$$\mathbf{u}_k(\varphi) = B_k^{-1}[U_k(\mathrm{Re}^{\mathrm{i}\varphi}) + \mathrm{i}U_k^*(\mathrm{Re}^{\mathrm{i}\varphi})], \quad k = 1, \ldots, |m| - 1,$$

where

$$B_k = \frac{1}{2\pi}\int_O^{2\pi} \mathrm{e}^{\mathrm{i}(m+1)\varphi}\Big[U_k(\mathrm{Re}^{\mathrm{i}\varphi}) + \mathrm{i}U_k^*(\mathrm{Re}^{\mathrm{i}\varphi})]h(\varphi)\,\mathrm{d}\varphi,$$

$v_O(\varphi) = 1, \quad v_k(\varphi) = \mathrm{Re}(B_k \mathrm{e}^{\mathrm{i}k\varphi}), \quad k = 1, \ldots, |m| - 1, \quad v_{|m|}(\varphi) = \mathrm{Re}(\mathrm{i}\mathcal{X}\mathrm{e}^{\mathrm{i}|m|\varphi})$, where $\mathcal{X} = 1$ for $\lambda > 0; \mathcal{X} = \mathrm{i}$ for $\lambda < 0$ and $\mathcal{X} = \mathrm{i}(|\lambda| - \lambda)$ for $\mathrm{Im}\,\lambda \neq 0$.

The second property is that

$$\frac{1}{2\pi}\int_O^{2\pi} \mathrm{e}^{\mathrm{i}(m+1)\varphi}\mathbf{u}_p(\varphi)\nu_p(\varphi)\,\mathrm{d}\varphi, = \delta_{pq} \quad (p, q = 0, 1, \ldots, |m|) \tag{9.40}$$

where $\delta_{pq} = O$ for $p \neq q$ and $\delta_{pq} = -1$ for $p = q$.

Now we assume that $\Phi(z)$ satisfies the boundary condition

$$\mathrm{Re}[z^{-m}\Phi] = h^*(z) \quad \text{on } |z| = R, \tag{9.41}$$

where

$$h^*(z) = h(z) + B_O v_O(\varphi) + B_{|m|}v_{|m|}(\varphi) + \sum_{k=1}^{|m|-1} v_k(\varphi),$$

the constants B_p (for indices $p = O$ and $p = |m|$) being defined by the formula

$$B_p = \frac{1}{2\pi\mathrm{i}}\int_O^{2\pi} \mathrm{e}^{\mathrm{i}(m+1)\varphi}U_p(Re^{\mathrm{i}\varphi})h(\varphi)\,\mathrm{d}\varphi.$$

From Theorem 9.3 and relation (9.10) the next result follows.

Consequence 9.1 *The inhomogeneous boundary problem* (9.1), (9.41) *for* $m < O$ *has a unique solution.*

In other words, $h^*(z)$ automatically satisfies the solvability conditions (9.19) or (9.36) which Theorem 9.3 treats about.

10 Conjugation problem for solutions of the model equation

We denote the disk $|z| < R$ by G^+, the circle $|z| < R$ by Γ and the domain $|z| > R$ by G^-. Let Φ^+ be a function from the class $C(G^+ + \Gamma) \cap C_{\bar{z}}(G^+ - 0)$ which is a solution of the model equation (9.1). We also consider the analytic function $\Phi^-(z)$, $z \in G^-$, bounded at $z = \infty$ and continuously extendable on Γ.

Problem *It is necessary to determine a pair of functions* $\Phi^+(z)$, $z \in G^+$, *and* $\Phi^-(z), z \in G^-$, *connected on* Γ *by the relation*

$$\Phi^+(z) = z^m \Phi^-(z) + g(z), \quad z \in \Gamma, \tag{10.1}$$

where m *is an integer and* $g(z)$ *is a given complex function expandable in absolutely and uniformly convergent Fourier series:*

$$g(Re^{i\varphi}) = \sum_{-\infty}^{\infty} g_k e^{ik\varphi}, \tag{10.2}$$

$$g_k = \frac{1}{2\pi} \int_O^{2\pi} g(Re^{i\varphi}) e^{-ik\varphi} \, d\varphi \quad (k = O, \pm 1, \ldots).$$

Similar to the previous section the solution to this problem is based on expressions for $\Phi^+(z)$ and $\Phi^-(z)$ in power series form:

$$\Phi^+(z) = \mathcal{X} a_O \left(\frac{r}{R}\right)^{|\lambda|} + \sum_{k=1}^{\infty} \left(\lambda a_k e^{ik\varphi} + P_{-k} \bar{a}_k e^{-ik\varphi}\right) \left(\frac{r}{R}\right)^{\mu_k}, \quad z < R \tag{10.3}$$

$$\Phi^-(z) = \sum_{k=O}^{\infty} \frac{b_k}{z^k}, \quad z > R.$$

The formulation of the final results essentially depends on the sign of m.

10.1 Theorem 10.1 *Let* $m \geq 1$. *Then a pair of functions* $\Phi^+(z)$ and $\Phi^-(z)$ *forming the solution of the homogeneous boundary problem* (10.1) ($g(z) = 0$) *contains* $2m + 1$ *arbitrary real constants and is given by the formulas:*

$$\Phi_O^+(z) = \mathcal{X} c_O \left(\frac{r}{R}\right)^{\mu_O} \sum_{k=1}^{m} \left(\lambda c_k e^{ik\varphi} + P_{-k}\bar{c}_k e^{-ik\varphi}\right)\left(\frac{r}{R}\right)^{\mu_k}, \tag{10.4}$$

$$R^m \Phi_O^-(z) = \mathcal{X} c_O \left(\frac{R}{z}\right)^m + \sum_{k=1}^{m} \lambda c_k \left(\frac{R}{z}\right)^{m-k} + P_{-k}\bar{c}_k \left(\frac{R}{z}\right)^{m+k},$$

where c_O *is an arbitrary real constant, and* c_k *is a complex constant.*

The inhomogeneous problem is always solvable.

Proof Substituting (10.2), (10.3) into (10.1) and equating coefficients of the same powers of $e^{i\varphi}$, we get a system of linear equations for determining the unknowns a_k, b_k:

$$\mathcal{X} a_O = b_m + g_O, \tag{10.5}$$

$$\lambda a_p = b_{m-p} R^p + g_p, \quad p = 1, \ldots, m, \tag{10.6}$$

$$\lambda a_k = g_k, \quad k \geq m + 1, \tag{10.7}$$

$$P_{-k}\bar{a}_k = b_{m+k} R^{-k} + g_{-k}, \quad k = 1, 2, \ldots \tag{10.8}$$

Assuming $a_O = c_O$ and $a_p = c_p, p = 1, \ldots, m$, where c_O is a real arbitrary constant and c_p is a complex one, from (10.5), (10.6), (10.8) we define

$$b_m = \mathcal{X} c_O - g_O,$$

$$b_{m-p} = \lambda c_p R^{-p} - g_p R^{-p},$$

$$b_{m+p} = P_{-k}\bar{c}_p R^p - g_{-p} R^p, \quad p = 1, \ldots, m.$$

From (10.7) we further get

$$a_k = \frac{1}{\lambda} g_k, \quad k \geq m + 1,$$

then we calculate the last group of unknown constants $b_{m+k}, k \geq m+1$, from (10.8):

$$b_{m+k} = \frac{1}{\bar{\lambda}} P_{-k}\bar{g}_k R^k - g_{-k} R^k.$$

Using this data in formulas (10.3), we present for convenience $\Phi^+(z)$ and $\Phi^-(z)$ as the sum of two series in the following way: the first series (with a finite number of terms) are composed from the terms containing the arbitrary constants c_O, c_k; the second series are constructed from the terms which do not contain these types of constants. The first series were written out in Theorem 10.1 under (10.4) and obviously give a complete aggregate of non–trivial solutions for the homogeneous conjugation problem. The second series have the forms:

$$\hat{\Phi}^+(z) = \sum_{k=m+1}^{\infty}\left(g_k e^{ik\varphi} + \frac{P_{-k}}{\bar{\lambda}}\bar{g}_k e^{-ik\varphi}\right)\left(\frac{r}{R}\right)^{\mu_k}, \quad |z| \leq R.$$

$$R^m\hat{\Phi}_O^-(z) = -\sum_{k=O}^{m} g_{m-k}\left(\frac{R}{z}\right)^k - \sum_{k=m+1}^{2m} g_{-(s-m)}\left(\frac{R}{z}\right)^k$$
$$+ \sum_{k=2m+1}^{\infty}\left(\frac{P_{-(k-m)}}{\bar{\lambda}}\bar{g}_{k-m} - g_{m-k}\right)\left(\frac{R}{z}\right)^k, \quad |z| \geq R.$$

The uniform convergence of these series in the above domains are an obvious consequence of the absolute convergence of the Fourier series for the function $g(z), |z| = R$. It is also clear that the pair $\hat{\Phi}^+(z)$ and $\hat{\Phi}^-(z)$ forms a partial solution for the inhomogeneous conjugation problem. Theorem 10.1 is proved.

10.2 We pass over to the conjugation problem in the case $m < O$.

Theorem 10.2 *Let $m < O$. Then the homogeneous problem ($g(z) = O$) has only the zero solution, and to solve the inhomogeneous problem it is necessary and sufficient that $2|m| - 1$ real constants satisfy*

$$\mathrm{Im}\left(\frac{1}{\mathcal{X}}g_O\right) = O, \quad \bar{\lambda}g_{-k} - P_{-k}\bar{g}_k = O, \quad (k = 1, \dots, |m| - 1). \tag{10.9}$$

Proof Taking $m = -|m| < O$ into account and introducing (10.2), (10.3) into boundary condition (10.1) we derive to the following system

$$\mathcal{X}a_O = g_O, \tag{10.10}$$

$$\lambda a_k = g_k, \quad k = 1, 2, \dots, \tag{10.11}$$

$$P_{-s}\bar{a}_s = g_{-s}, \quad s = 1, \dots, |m| - 1, \tag{10.12}$$

$$P_{-k}\bar{a}_k = b_{k-|m|}R^{-k} + g_{-k}, \quad k \geq m. \tag{10.13}$$

Dividing relation (10.10) by $\mathcal{X}$ and then separating out its real part, we get

$$a_O = \mathrm{Re}\left(\frac{1}{\mathcal{X}} g_O\right). \tag{10.14}$$

As to the imaginary part, it restricts g_O, which characterizes one of the solvability conditions of the problem and was written out in Theorem 10.2.

From (10.11) we have

$$a_k = \frac{1}{\lambda} g_k, \quad k = 1, 2, \ldots, \tag{10.15}$$

If one introduces the value a_k into (10.12) and (10.13), then from (10.12) we shall get $|m| - 1$ complex solvability conditions which are contained in (10.9), and from (10.13) determine the unknown constants b_k:

$$R^{-(k+|m|)} b_k = \frac{P_{-(k+|m|)}}{\bar{\lambda}} \bar{g}_{k+|m|} - g_{-(k+|m|)}. \tag{10.16}$$

In particular, it follows from (10.14)–(10.16) that all the unknown constants a_k, b_k will be equal to zero if $g(z) \equiv 0$.

For $g(z) \not\equiv 0$, under the assumption that (10.9) are fulfilled, the pair of functions $\Phi^{\pm}(z)$, which gives the unique solution of the conjugation problem, is determined by the equalities:

$$\Phi^+(z) = \mathcal{X}\mathrm{Re}\left(\frac{1}{\mathcal{X}} g_O\right)\left(\frac{r}{R}\right)^{|\lambda|} + \sum_{k=1}^{\infty}\left(g_k e^{ik\varphi} + \frac{P_{-k}}{\bar{\lambda}} \bar{g}_k e^{-ik\varphi}\right)\left(\frac{r}{R}\right)^{\mu_k},$$

$$\Phi^-(z) = R^{|m|} \sum_{k=O}^{\infty}\left(\frac{P_{-(k+|m|)}}{\bar{\lambda}} \bar{g}_{k+|m|} - g_{-(k+|m|)}\right)\left(\frac{R}{z}\right)^k.$$

The uniform convergence of these series issues from the absolute convergence of the Fourier series for $g(z)$ as in Theorem 10.1.

Theorem 10.2 is proved.

Let us consider instead of (10.1) the boundary condition in the form

$$\Phi^+(z) = z^m \Phi^-(z) + g^*(z), \quad |z| = R, \tag{10.17}$$

where

$$g^*(z) = g(z) - \mathrm{i}\,\mathcal{X}\,\mathrm{Im}\left(\frac{g_O}{\mathcal{X}}\right) - \sum_{k=1}^{|m|-1}\left(g_k - \frac{P_{-k}}{\bar{\lambda}}\bar{g}_k\right)\mathrm{e}^{-\mathrm{i}k\varphi}.$$

From Theorem 10.2 one deduces

Consequence 10.1 *For* $m < O$ *the inhomogeneous conjugation problem with boundary condition* (10.17) *has a unique solution.*

11 Additions to Chapter 2

This section is devoted to some results of the theory of the model equation, but which will not be utilized in this book. Therefore their explanation will be presented in a concise form.

11.1 The behaviour of solutions in a neighbourhood of an isolated singular point of coefficients of equations has been investigated by G. Nazirov [40] and N.K. Bliev [10–12] for the equation of the following type

$$r\mathrm{e}^{\mathrm{i}n\varphi}\partial_{\bar{z}}w + a(z)w + b(z)\bar{w} = 0, \tag{11.1}$$

in which $z = r\mathrm{e}^{\mathrm{i}\varphi}, n = -1, 1; a(z)$ and $b(z)$ are analytic functions of z and $\bar{z}$ in a certain neighbourhood of the point $z = O$:

$$a(z) = \sum_{k,s=O}^{\infty} a_{k,s}z^k\bar{z}^s, \qquad b(z) = \sum_{k,s=O}^{\infty} b_{k,s}z^k\bar{z}^s.$$

The existence of the solutions

$$w(z) = |z|^{2\alpha}\sum_{k,s=O}^{\infty} c_{k,s}z^k\bar{z}^s,$$

has been proved in [40] for $a(z) \equiv O$ and $b(O) = b_{O,O} \neq 0$. Here α is calculated from $|\alpha + 1|^2 - |b(O)|^2 = O$ for $n = 1$ and $\alpha = |b(0)|\mathrm{e}^{\mathrm{i}\gamma}$; γ is an arbitrary angle with $n = -1$.

In [10–12] necessary and sufficient conditions for the existence of analytic solutions with respect to z and $\bar{z}$ in a neighbourhood of $z = 0$ have been established. Thus for equation (11.1) with $n = -1$ the whole question is reduced to the existence of the solutions to the algebraic equation.

$$(a_{O,O} - l)[\bar{a}_{O,O} - (m - l)] = b_{O,O}^2.$$

in integer l and $m, 0 \le l \le [m/2]$.

11.2 For investigating different problems, connected with the model equation, certain advantages can be taken of all kinds of representations of its solutions through analytic functions. One of these is extracted from the integral equation

$$\Phi(z) + \frac{\lambda}{2\pi} \iint\limits_{|z|\le R} \frac{\overline{\Phi(\zeta)}}{\bar{\zeta}(\zeta - z)}\, d\xi\, d\eta = \Psi(z), \quad |z| \le R. \tag{11.2}$$

If one supposes that $\Psi(z)$ is an analytic function of the complex variable z in the domain $|z| \le R$ then the unknown function $\Phi(z)$, as follows from [69], will be a solution of the model equation (9.1).

In L. Mikhailov's paper [37] the solvability of equation (11.2) has been obtained provided that $\Phi(z)$ and $\Psi(z)$ in a disk $|z| \le R$ belong to one of the class $C_\beta, M_\beta, L^p_{\beta-2/p}, 0 < \beta < 1$ (a function $f(z) \in C_\beta$ if it is expressed in the form $f(z) = |z|^{-\beta} f_O(z)$, while $f_O(z)$ belongs to the class C of continuous functions; M_β and $L^p_{\beta-2/p}$ are defined in a similar way [36]. It has been proved that equation (11.2) for $\Psi(z) \equiv O$ and any values λ has only a zero solution. For $\Psi \ne O$ the solvability picture depends on $|\lambda|$ and β. If $|\lambda| < \beta$, then (11.2) has a solution for any $\Psi(z)$; if $|\lambda| > \beta$ then a solution exists if and only if $\Psi(z)$ satisfies the condition

$$\iint\limits_{|z|\le R} \left\{ \Psi(z) - \frac{\lambda}{|\lambda|} \Psi(z) \right\} |\zeta|^{|\lambda|-2}\, d\xi\, d\eta = 0. \tag{11.3}$$

An efficient supplement to this result has been worked out in [58] where equation (11.2) has been solved in explicit form. To formulate this result we introduce the functions Ω_1^* and Ω_2^* in the domain $|z|, |\zeta| \le R$:

$$\Omega_1^* = \frac{\delta}{2}\left(\frac{\rho}{r}\right)^{|\lambda|} + \sum_{k=1}^{\infty} \frac{P_{-k}}{2\mu_k}\left(e^{ik(\varphi-\gamma)} - e^{-ik(\varphi-\gamma)}\right)\left(\frac{\rho r}{R^2}\right)^{\mu_k},$$

$$\Omega_2^* = -\frac{\delta\lambda}{2|\lambda|}\left(\frac{\rho}{r}\right)^{|\lambda|} - \sum_{k=1}^{\infty} \frac{|\lambda|^2 e^{ik(\varphi-\gamma)} - P_{-k}^2 e^{-ik(\varphi-\gamma)}}{2\bar{\lambda}\mu_k}\left(\frac{\rho r}{R^2}\right)^{\mu_k}.$$

In the above case we use familiar notation: $z = re^{i\varphi}, \zeta = \rho e^{i\gamma}, \mu_k^2 = |\lambda|^2 + k^2, P_{-k} = \mu_k - k$; in addition $\delta = 1$ for $|\lambda| < \beta$ and $\delta = 0$ for $|\lambda| > \beta$.

Immediate verification establishes that these functions together with Ω_1 and Ω_2 defined by formulas (1.12), ch. 1, satisfy the relations:

$$-\frac{\lambda}{2\pi}\iint\limits_{|\zeta_1|\le R}\frac{\overline{\Omega_1(\zeta_1,\zeta)}+\overline{\Omega_1^*(\zeta_1,\zeta)}}{\bar{\zeta}_1(\zeta_1-z)}\,d\xi_1\,d\eta_1=\Omega_2(z,\zeta)+\Omega_2^*(z,\zeta)-\frac{\lambda(1-\delta)}{2|\lambda|}\left(\frac{\rho}{R}\right)^{|\lambda|},$$

$$-\frac{\lambda}{2\pi}\iint\limits_{|\zeta_1|\le R}\frac{\Omega_2(\zeta_1,\zeta)+\Omega_2^*(\zeta_1,\zeta)}{\bar{\zeta}_1(\zeta_1-z)}\,d\xi_1\,d\eta_1=\Omega_1(z,\zeta)+\Omega_1^*(z,\zeta)-\frac{\zeta}{\zeta-z}+\frac{1-\delta}{2}\left(\frac{\rho}{R}\right)^{|\lambda|}.$$

(11.4)

Theorem 11.1 *Let $\Psi(z)$ belong to the class $C_\beta, 0<\beta<1$, for $|z|\le R$ and in the case of $|\lambda|>\beta$ let it satisfy condition* (11.3). *Then $\Phi(z)$ given by the relation*

$$\Phi(z)=\Psi(z)-\frac{1}{2\pi}\iint\limits_{|\zeta|\le R}\left\{\lambda[\Omega_1(z,\zeta)+\Omega_1^*(z,\zeta)]\overline{\Psi(\zeta)}+\bar{\lambda}[\Omega_2(z,\zeta)+\Omega_2^*(z,\zeta)]\Psi(\zeta)\right\}\frac{d\xi\,d\eta}{|\zeta|^2},$$

(11.5)

is a solution of the integral equation (11.2) *from the C_β class.*

To prove this result the right side of (11.5) should be introduced into equation (11.2) instead of $\Phi(z)$. Then we should use (11.4).

Formula (11.5) gives the representation solutions $\Phi(z)$ of the model equation through the analytic functions $\Psi(z)$.

Chapter 3
The general equation

In this chapter the subject of our investigation will be the equation

$$\partial_{\bar{z}} w - \frac{b(z)}{2\bar{z}} \bar{w} = F(z), \quad z \in G,$$

in which the singular point $z = 0$ is situated inside the domain G, $w(z)$ is an unknown function, $F(z)$ and $b(z)$ are given functions, while $F(z) \in C(G - O) \cap L_p(\bar{G}), p > 2$, and $b(z) \in C(G), b(0) \neq 0$ and

$$|b(z) - b(0)| < M|z|^{\alpha}, \quad a > 0,$$

at least for z, belonging to a small neighbourhood of $z = O$.

This equation extends the model equation, analysed in the previous chapter, and therefore will be called *the general equation*. Its study is developed on the basis of the theory of the model equation. The integral equation (4.2), ch. 1, establishing a relation between sets of the general equation and the model equation, will be utilized as the main tool, by means of which some of the results, received for the model object, will be transformed into the object of a more complicated nature. It is in this way that we reveal the analogy of the asymptotic behaviour in a neighbourhood of $z = O$ of continuous solutions of the general and model equations, as well as their solvability of boundary value problems [55, 56, 59, 60, 65].

1 Behaviour of solutions at a singular point

We consider the equation

$$\partial_{\bar{z}} w - \frac{b(z)}{2\bar{z}} \bar{w} = O, \quad z \in G, \tag{1.1}$$

where $b(z)$ is subordinated to the above conditions. Let $w(z)$ be its solution from the class $C(\bar{G}) \cap C_{\bar{z}}(G - O)$. We study the asymptotic behaviour $w(z)$ in a neighbourhood of the point $z = O$, using the solution represented in the following form:

$$w(z) = \Phi(z) - \frac{1}{\pi} \iint\limits_G \left[\frac{\Omega_1(z, \zeta)}{\zeta} f(\zeta) + \frac{\Omega_2(z, \zeta)}{\bar{\zeta}} \overline{f(\zeta)} \right] d\xi \, d\eta, \tag{1.2}$$

where

$$f(\zeta) = \frac{b(\zeta) - b(0)}{2\bar{\zeta}}\overline{w(\zeta)} \tag{1.3}$$

and $\Phi(z)$ is a continuous solution of the model equation, determined uniquely by the given $w(z)$:

$$\Phi(z) = \frac{1}{2\pi i}\int_\Gamma \frac{\Omega_1(z,\zeta)}{\zeta}w(\zeta)\,d\zeta - \frac{\Omega_2(z,\zeta)}{\bar{\zeta}}\overline{w(\zeta)}\,\overline{d\zeta}.$$

This solution representation is the result of using formula (2.7), ch.2, in equation (1.1), which it is necessary to rewrite beforehand in (2.6), ch.2, employing notation (1.3) and $\lambda = b(O)$.

It is fairly clear that we can restrict ourselves to considering a disk of radius R with the centre at the point $z = O$ as the domain G. Then $\Phi(z)$ can be expressed by a power series as was done in §4, ch. 2:

$$\Phi(z) = \left(a_O + \frac{\lambda}{|\lambda|}\overline{a_O}\right)r^{|\lambda|} + \sum_{k=1}^{\infty}\left(\frac{P_k a_k + \lambda\overline{a_{-k}}}{\mu_k}\,e^{ik\varphi} + \frac{P_{-k}a_{-k} + \lambda\overline{a_k}}{\mu_k}\,e^{-ik\varphi}\right)r^{\mu_k}, \tag{1.4}$$

where $\lambda = b(0)$ and

$$a_k = \frac{R^{-\mu_k}}{4\pi}\int_O^{2\pi} w(Re^{i\theta})e^{-ik\theta}\,d\theta, \quad k = 0, \pm 1, \ldots. \tag{1.5}$$

Theorem 1.1 *Let $w(z)$ be a solution of equation (1.1) from $C(\bar{G}) \cap C_{\bar{z}}(G - 0)$. Then for $w(z)$ the following relation is valid:*

$$w(z) = \left(\beta_O + \frac{\lambda}{|\lambda|}\beta_O\right)r^{|\lambda|} + O\left(r^{|\lambda| + a_0}\right), \quad r \to 0. \tag{1.6}$$

If $w(z)$ is subordinated to k conditions of the form

$$P_s a_s + \lambda\bar{a}_{-s} = \frac{1}{2\pi}\iint\limits_G \left[\frac{P_s}{\zeta}f(\zeta) + \frac{\lambda}{\bar{\zeta}}\overline{f(\zeta)}\right]e^{-is\gamma}\frac{d\xi\,d\eta}{\rho^{\mu_s}} \quad (s = 0, 1, \ldots, k-1), \tag{1.7}$$

where $a_s, a_{-s}, f(\zeta)$ are defined by equalities (1.5) and (1.3), then the asymptotic behaviour of $w(z)$ in a neighbourhood of the point $z = O$ will be characterized by the formula

$$w(z) = \left(\frac{P_k\beta_k + \lambda\bar{\beta}_{-k}}{\mu_k}\,e^{ik\varphi} + \frac{P_{-k}\beta_{-k} + \lambda\bar{\beta}_k}{\mu_k}e^{-ik\varphi}\right)r^{\mu_k} + O\left(r^{\mu_k + a_k}\right), \quad r \to 0. \tag{1.8}$$

Here $a_k(k = O, 1, \ldots)$ is some positive number and

$$\beta_k = a_k - \frac{1}{2\pi} \iint\limits_G \frac{f(\zeta)}{\zeta} \mathrm{e}^{-ik\gamma} \frac{d\xi\, d\eta}{\rho^{\mu_k}}, \quad k = 0, \pm 1, \ldots. \tag{1.9}$$

Proof We assume that $\alpha \leq |\lambda|$ (α is an exponent in the Holder inequality for the coefficient $b(z)$; see the introduction to this chapter. First of all we get the relation

$$w(z) = O\Big(|z|^{\alpha - \varepsilon}\Big), \quad z \to 0, \tag{1.10}$$

where ε is an arbitrary small positive number. From (1.2) we have

$$\frac{w(z)}{|z|^{\alpha-\varepsilon}} = \frac{\Phi(z)}{|z|^{\alpha-\varepsilon}} + J_1(z), \tag{1.11}$$

where

$$J_1(z) = -\frac{1}{\pi} \iint\limits_G \left[\frac{\Omega_1^{(1)}(z,\zeta)}{\zeta} f^{(1)}(\zeta) + \frac{\Omega_2^{(1)}(z,\zeta)}{\bar{\zeta}} \overline{f^{(1)}(\zeta)}\right] d\xi\, d\eta,$$

while

$$f^{(1)}(\zeta) = \frac{f(\zeta)}{|\zeta|^{\alpha-\varepsilon}}, \qquad \Omega_k^{(1)}(z,\zeta) = \left|\frac{\zeta}{z}\right|^{\alpha-\varepsilon} \Omega_k(z,\zeta), \quad k = 1, 2.$$

We verify that the right side of (1.11) is bounded at $z = 0$. The boundedness of the first term issues from (1.4) by virtue of $\alpha \leq |\lambda|$. It remains to consider $J_1(z)$. Representing $\Omega_1^{(1)}(z,\zeta)$ in the form

$$\Omega_1^{(1)}(z,\zeta) = \frac{\zeta}{\zeta - z} + \Omega_1^{(1)o}(z,\zeta),$$

we have

$$J_1(z) = -\frac{1}{\pi} \iint\limits_G \frac{f^{(1)}(\zeta)}{\zeta - z} d\xi\, d\eta - \frac{1}{\pi} \iint\limits_G \left[\frac{\Omega_1^{(1)o}(z,\zeta)}{\zeta} f^{(1)}(\zeta) + \frac{\Omega_2^{(1)}(z,\zeta)}{\bar{\zeta}} \overline{f^{(1)}(\zeta)}\right] d\xi\, d\eta.$$

On the basis of this formula the boundedness of $J_1(z)$ is established by the scheme in which the upper bound for the function $h(z)$ has been obtained in Lemma 3.1 ch.1. We note only that $f^{(1)}(\zeta) \in L_p(\bar{G})$, where p is subordinated to the inequalities $2 < p < 2/(1 - \varepsilon)$.

Now (1.10) follows from the boundedness of the right side of (1.11). Using this result for the further correction of the order of zero of $w(z)$ at the point $z = O$, we write down (1.2) in the form

$$\frac{w(z)}{|z|^{2\alpha-2\varepsilon}} = \frac{\Phi(z)}{|z|^{2\alpha-2\varepsilon}} + J_2(z) \tag{1.12}$$

where

$$J_2(z) = -\frac{1}{\pi}\iint_G \left[\frac{\Omega_1^{(2)}(z,\zeta)}{\zeta} f^{(2)}(\zeta) + \frac{\Omega_2^{(2)}(z,\zeta)}{\bar\zeta}\overline{f^{(2)}(\zeta)}\right] \mathrm{d}\xi\, \mathrm{d}\eta,$$

while

$$f^{(2)}(\zeta) = \frac{f(\zeta)}{|\zeta|^{2\alpha-2\varepsilon}}, \qquad \Omega_k^{(2)}(z,\zeta) = \left|\frac{\zeta}{z}\right|^{2\alpha-2\varepsilon} \Omega_k(z,\zeta), \quad k = 1,2.$$

Establishing the boundedness of the right side of (1.12) in a similar way, we arrive at the equality $w(z) = O(|z|^{2\alpha-2\varepsilon}), z \to 0$. Repeating this procedure as many times as is necessary, we get

$$w(z) = O(|z|^{|\lambda|}), \quad z \to 0.$$

Now we return to the relation (1.2) and change it into the following form

$$w(z) = \left(\beta_O + \frac{\lambda}{|\lambda|}\bar\beta_O\right) r^{|\lambda|} + \tilde\Phi(z) - \frac{1}{\pi}\iint_G \left[\frac{\tilde\Omega_1(z,\zeta)}{\zeta} f(\zeta) + \frac{\tilde\Omega_2(z,\zeta)}{\bar\zeta}\overline{f(\zeta)}\right] \mathrm{d}\xi\, \mathrm{d}\eta, \tag{1.13}$$

where β_O is determined by the equality (1.9),

$$\tilde\Phi(z) = \sum_{k=1}^{\infty}\left(\frac{P_k a_k + \lambda\overline{a_{-k}}}{\mu_k}\mathrm{e}^{\mathrm{i}k\varphi} + \frac{P_{-k}a_{-k} + \lambda\overline{a_k}}{\mu_k}\mathrm{e}^{-\mathrm{i}k\varphi}\right) r^{\mu_k},$$

and in addition, cf. (1.12), ch. 1,

$$\tilde\Omega_1 = \begin{cases} \displaystyle\sum_{k=1}^{\infty} \frac{P_k \mathrm{e}^{\mathrm{i}k(\varphi-\gamma)} + P_{-k}\mathrm{e}^{-\mathrm{i}k(\varphi-\gamma)}}{2\mu_k}\left(\frac{r}{\rho}\right)^{\mu_k}, |z| < |\zeta|, \\ \displaystyle -\frac{1}{2}\left(\frac{r}{\rho}\right)^{|\lambda|} - \frac{1}{2}\left(\frac{\rho}{r}\right)^{|\lambda|} - \sum_{k=1}^{\infty} \frac{P_{-k}\mathrm{e}^{\mathrm{i}k(\varphi-\gamma)} + P_k\mathrm{e}^{-\mathrm{i}k(\varphi-\gamma)}}{2\mu_k}\left(\frac{\rho}{r}\right)^{\mu_k}, |z| > |\zeta|, \end{cases}$$

$$\tilde{\Omega}_2 = \begin{cases} \displaystyle\sum_{k=1}^{\infty} \frac{\lambda}{\mu_k} \cos k(\varphi - \gamma)\left(\frac{r}{\rho}\right)^{\mu_k}, |z| < |\zeta|, \\ \displaystyle -\frac{\lambda}{2|\lambda|}\left(\frac{r}{\rho}\right)^{|\lambda|} + \frac{\lambda}{2|\lambda|}\left(\frac{\rho}{r}\right)^{|\lambda|} + \sum_{k=1}^{\infty} \frac{\lambda}{\mu_k} \cos k(\varphi - \gamma)\left(\frac{\rho}{r}\right)^{\mu_k}, |z| > |\zeta|. \end{cases}$$

For deducing (1.6) from (1.13) it remains to check that the second and third terms in formula (1.13) have a zero of order $|\lambda| + a_O$ at the point $z = O$. For the second one, i.e. for $\tilde{\Phi}(z)$, it is obvious. The test of the third one is similar to that in the proof of (1.10) (you may see it in the text below).

Thus, we have obtained (1.6). The validity of (1.8) is established by the induction method. Omitting the redundant awkwardness, we prove it for $k = 1$.

First of all we must assume

$$\beta_O + \frac{\lambda}{|\lambda|}\bar{\beta}_O = 0, \tag{1.14}$$

which is equivalent to condition (1.7) for $s = O$ thanks to the notation (1.9). Taking (1.14) into account, from (1.13) we have

$$w(z) = \tilde{\Phi}(z) - \frac{1}{\pi} \iint_G \left[\frac{\tilde{\Omega}_1(z,\zeta)}{\zeta} f(\zeta) + \frac{\tilde{\Omega}_2(z,\zeta)}{\bar{\zeta}} \overline{f(\zeta)}\right] d\xi\, d\eta. \tag{1.15}$$

If $|\lambda| + a \le \mu_1$, then we can rewrite (1.15) in the following form:

$$\frac{w(z)}{|z|^{|\lambda|+\alpha-\varepsilon}} = \frac{\tilde{\Phi}(z)}{|z|^{|\lambda|+\alpha-\varepsilon}} + \tilde{J}_1(z), \tag{1.16}$$

where

$$\tilde{J}_1(z) = -\frac{1}{\pi} \iint_G \left[\frac{\tilde{\Omega}_1^{(1)}(z,\zeta)}{\zeta} \tilde{f}^{(1)}(\zeta) + \frac{\tilde{\Omega}_2^{(1)}(z,\zeta)}{\bar{\zeta}} \overline{\tilde{f}^{(1)}}(\zeta)\right] d\xi\, d\eta,$$

while

$$\tilde{f}^{(1)}(\zeta) = \frac{f(\zeta)}{|\zeta|^{|\lambda|+\alpha-\varepsilon}} \quad \text{and} \quad \tilde{\Omega}_k^{(1)}(z,\zeta) = \left|\frac{\zeta}{z}\right|^{|\lambda|+\alpha-\varepsilon} \tilde{\Omega}_k(z,\zeta), \quad k = 1,2.$$

And again in a similar way the bounded right side of (1.16) is established. This leads to the relation $w(z) = O(|z|^{|\lambda|+\alpha-\varepsilon})$, $z \to 0$. We further get $w(z) = O(|z|^{|\lambda|+2\alpha-2\varepsilon})$, $z \to O$ etc., and after a finite number of steps we have $w(z) = O(|z|^{\mu_1})$, $z \to 0$.

Now we return to (1.15) and rewrite this formula in the form

$$
\begin{aligned}
w(z) =& \left(\frac{P_1\beta_1 + \lambda\bar{\beta}_{-1}}{\mu_1} \mathrm{e}^{\mathrm{i}\varphi} + \frac{P_{-1}\beta_{-1} + \lambda\bar{\beta}_1}{\mu_1} \mathrm{e}^{-\mathrm{i}\varphi} \right) r^{\mu_1} \\
&+ \tilde{\Phi}(z) - \frac{1}{\pi} \iint\limits_G \left[\frac{\tilde{\tilde{\Omega}}_1(z,\zeta)}{\zeta} f(\zeta) + \frac{\tilde{\tilde{\Omega}}_2(z,\zeta)}{\bar{\zeta}} \overline{f(\zeta)} \right] \mathrm{d}\xi\, \mathrm{d}\eta,
\end{aligned} \tag{1.17}
$$

where β_1 and β_{-1} are defined by equality (1.9):

$$
\tilde{\tilde{\Phi}}(z) = \sum_{k=2}^{\infty} \left(\frac{P_k a_k + \lambda \overline{a_{-k}}}{\mu_k} \mathrm{e}^{\mathrm{i}k\varphi} + \frac{P_{-k} a_{-k} + \lambda a_k}{\mu_k} \mathrm{e}^{-\mathrm{i}k\varphi} \right) r^{\mu_k},
$$

and

$$
\tilde{\tilde{\Omega}}_1 = \begin{cases}
\displaystyle \sum_{k=2}^{\infty} \frac{P_k \mathrm{e}^{\mathrm{i}k(\varphi-\gamma)} + P_{-k}\mathrm{e}^{-\mathrm{i}k(\varphi-\gamma)}}{2\mu_k} \left(\frac{r}{\rho}\right)^{\mu_k}, |z| < |\zeta|, \\
\displaystyle -\frac{1}{2}\left(\frac{r}{\rho}\right)^{|\lambda|} - \frac{P_1 \mathrm{e}^{\mathrm{i}(\varphi-\gamma)} + P_{-1}\mathrm{e}^{-\mathrm{i}(\varphi-\gamma)}}{2\mu_1} \left(\frac{r}{\rho}\right)^{\mu_1} \quad |z| > |\zeta|, \\
\displaystyle -\frac{1}{2}\left(\frac{\rho}{r}\right)^{|\lambda|} - \sum_{k=1}^{\infty} \frac{P_{-k} \mathrm{e}^{\mathrm{i}k(\varphi-\gamma)} + P_{k}\mathrm{e}^{-\mathrm{i}k(\varphi-\gamma)}}{2\mu_k} \left(\frac{\rho}{r}\right)^{\mu_k}
\end{cases}
$$

$$
\tilde{\tilde{\Omega}}_2 = \begin{cases}
\displaystyle \sum_{k=2}^{\infty} \frac{\lambda}{\mu_k} \cos k(\varphi-\gamma) \left(\frac{r}{\rho}\right)^{\mu_k}, |z| < |\zeta|, \\
\displaystyle -\frac{\lambda}{2|\lambda|}\left(\frac{r}{\rho}\right)^{|\lambda|} - \frac{\lambda}{\mu_1} \cos(\varphi-\gamma) \left(\frac{r}{\rho}\right)^{\mu_1} \quad |z| > |\zeta|. \\
\displaystyle +\frac{\lambda}{2|\lambda|}\left(\frac{\rho}{r}\right)^{|\lambda|} + \sum_{k=1}^{\infty} \frac{\lambda}{\mu_k} \cos k(\varphi-\gamma) \left(\frac{\rho}{r}\right)^{\mu_k}
\end{cases}
$$

But again one verifies that terms 2 and 3 in formula (1.17) have a zero of order $\mu_1 + a_1$, where a_1 is a positive number. Theorem 1.1 is proved.

2 Solutions bounded on the plane

We return to equation (1.1). According to [69] any solution of equation (1.1) can be expressed in the form

$$
w(z) = \Psi(z)\mathrm{e}^{\omega(z)}, \tag{2.1}
$$

where $\Psi(z)$ is an analytic function in a domain G and

$$\omega(z) = -\frac{1}{2\pi}\iint\limits_G \frac{b(\zeta)}{\bar{\zeta}(\zeta - z)}\frac{\overline{w(\zeta)}}{w(\zeta)}\,d\xi\,d\eta. \tag{2.2}$$

As before we assume that $w(z) \in C(G) \cap C_{\bar{z}}(G - O)$ and $b(z)$ satisfies the conditions mentioned in the introduction to this chapter.

In the present section on the basis of Theorem 1.1 we shall analyse the behaviour of $\omega(z)$ at the point $z = O$, then characterize the possible structure of zeros for the analytic function $\Psi(z)$ at the same point, and finally prove a variant of Liouville's theorem for the solutions of equation (1.1).

2.1 Lemma 2.1 *The behaviour of $\omega(z)$ in a neighbourhood of $z = O$ is characterized by the relation*

$$\omega(z) = P_{-k}\ln r + \ln\left(1 + \frac{P_{-k}}{\lambda}\frac{\bar{B}_k}{B_k}e^{-2ik\varphi}\right) + \omega_O^*(z), \tag{2.3}$$

where k can take the values $0, 1, 2, \ldots; \lambda = b(O)$, $\mu_k^2 = |\lambda|^2 + k^2$, $P_{-k} = \mu_k - k, B_k$ is a constant, and $\omega_O^*(z)$ is a continuous function

Proof We write $\omega(z)$ in the following form:

$$\omega(z) = -\frac{\lambda}{2\pi}\iint\limits_G \frac{\overline{w(\zeta)}d\xi\,d\eta}{w(\zeta)\bar{\zeta}(\zeta - z)} + \omega_O(z), \tag{2.4}$$

where $\lambda = b(0)$ and

$$\omega_O(z) = -\frac{1}{2\pi}\iint\limits_G \frac{[b(\zeta) - b(0)]\overline{w(\zeta)}}{\bar{\zeta}(\zeta - z)w(\zeta)}\,d\xi\,d\eta$$

is a continuous function.

According to Theorem 1.1 the asymptotic behaviour of $w(z)$ in a neighbourhood of $z = O$ is defined by formula (1.8). For $k = O$ it contains (1.6). Using (1.8), we write the relation

$$\frac{\overline{w(\zeta)}}{w(\zeta)} = \frac{(P_k\bar{\beta}_k + \bar{\lambda}\beta_{-k})e^{-ik\varphi} + (P_{-k}\bar{\beta}_{-k} + \bar{\lambda}\beta_k)e^{ik\varphi}}{(P_{-k}\beta_{-k} + \lambda\bar{\beta}_k)e^{-ik\varphi} + (P_k\beta_k + \lambda\bar{\beta}_{-k})e^{ik\varphi}} + \mathcal{X}(z), \tag{2.5}$$

in which $\chi(z)$ is bounded measurable function and $\chi(z) = O(|z|^{\alpha_k}), z \to O, a_k > O$.

Let $B_k = P_{-k}\bar{\beta}_{-k} + \bar{\lambda}\beta_k$. Then $P_k\bar{\beta}_k + \bar{\lambda}\beta_{-k} = \bar{B}_k\bar{\lambda}/P_{-k}$. Rewriting (2.5) in this notation and substituting it into (2.4) we get

$$\omega(z) = -\frac{\lambda}{2\pi}\iint\limits_G \frac{\frac{\bar{\lambda}}{P_{-k}}\bar{B}_k e^{-ik\gamma} + B_k e^{ik\gamma}}{\bar{B}_k e^{-ik\gamma} + \frac{\lambda}{P_{-k}} B_k e^{ik\gamma}} \frac{d\xi\, d\eta}{\bar{\zeta}(\zeta - z)} + \omega_{OO}(z), \tag{2.6}$$

where $\gamma = \arg\zeta$ and $\omega_{OO}(\zeta) = \omega_O(\zeta) - \dfrac{\lambda}{2\pi}\displaystyle\iint\limits_G \frac{\chi(\zeta)\, d\xi\, d\eta}{\bar{\zeta}(\zeta - z)}$ is a continuous function. Without loss of generality we can assume that the domain G is the disk $|z| \leq R$. For the transformation of integral (2.6) we introduce a new complex variable σ by the equality $\zeta = z\sigma$. Applying polar coordinates $\zeta = \rho e^{i\gamma}, z = re^{i\varphi}, \sigma = \tau e^{i\alpha}$, we get $\rho = r\tau, \gamma = \varphi + \alpha,\ \bar{\zeta}(\zeta - z) = |z|^2\bar{\sigma}(\sigma - 1),\ d\xi\, d\eta = |z|^2\tau\, d\tau\, d\alpha$.

Therefore, (2.6) takes the form

$$\omega(z) = \omega_{OO}(z) - \frac{\lambda}{2\pi}\iint\limits_{\tau \leq Rr} \frac{\frac{\bar{\lambda}}{P_{-k}}\bar{B}_k e^{-ik(\varphi+\alpha)} + B_k e^{ik(\varphi+\alpha)}}{\bar{B}_k e^{-ik(\varphi+\alpha)} + \frac{\lambda}{P_{-k}} B_k e^{ik(\varphi+\alpha)}} \frac{d\tau\, d\alpha}{\tau - e^{-i\alpha}}.$$

Denoting this integral by J and introducing the complex variable $t = e^{i\alpha}$, after obvious transformation we get

$$J = -\frac{\lambda}{2\pi i}\int_0^{R/r} \frac{d\tau}{\tau}\int_{|t|=1} \frac{\frac{\bar{\lambda}}{P_{-k}}\frac{\bar{B}_k}{B_k} e^{-2ik\varphi} + t^{2k}}{\frac{\bar{B}_k}{B_k} e^{-2ik\varphi} + \frac{\lambda}{P_{-k}} t^{2k}} \frac{dt}{t - \frac{1}{t}}.$$

The subintegral expression is a one–valued analytic function with respect to t in the domain $|t| < 1$. Therefore for the calculation of the internal integral we can use the deduction theorem [34]. Taking into account the fact that the denominator of the subintegral expression has a simple zero at the point $t = I/\tau$ for $\tau > 1$ and $2k$ simple zeros at points $\beta_p (p = 0, 1, \ldots, 2k - 1)$:

$$\beta_p = \left(\frac{P_{-k}}{|\lambda|}\right)^{1/2k} e^{i(\varphi_p - \varphi)},$$

where $\varphi_p = \frac{1}{\pi}(\pi + 2\pi p - 2\arg B_k - \arg\lambda)$, we get

$$J = -\lambda\sum_{p=0}^{2k-1}\int_0^{R/r} \frac{\frac{\bar{\lambda}}{P_{-k}}\frac{\bar{B}_k}{B_k} e^{-2ik\varphi} + \beta_p^{2k}}{\frac{\bar{B}_k}{B_k} e^{-2ik\varphi} + \frac{\lambda}{P_{-k}}(2k+1)\beta_p^{2k} - \frac{2\lambda k}{P_{-k}}\frac{1}{\tau}\beta_p^{2k-1}} \frac{d\tau}{\tau}$$

$$-\lambda\int_1^{R/r}\frac{\frac{\lambda}{P_{-k}}\frac{\bar{B}_k}{B_k}e^{-2ik\varphi}+\left(\frac{1}{\tau}\right)^{2k}}{\frac{\bar{B}_k}{B_k}e^{-2ik\varphi}+\frac{\lambda}{P_{-k}}\left(\frac{1}{\tau}\right)^{2k}}\frac{d\tau}{\tau}. \tag{2.7}$$

With the notation

$$a=\frac{\bar{\lambda}}{P_{-k}}\frac{\bar{B}_k}{B_k}e^{-2ik\varphi},\qquad b=\frac{\bar{B}_k}{B_k}e^{-2ik\varphi},\qquad c=\frac{\lambda}{P_{-k}},\qquad \gamma_p=\alpha+\beta_p^{2k},$$

$$\delta_p=b+\frac{\lambda}{P_{-k}}(2k+1)\beta_p^{2k},\qquad c_p=\frac{2k}{P_{-k}}\beta_p^{2k-1}$$

the integration result of (2.7) gives

$$J=-\lambda\left(\frac{a}{b}+\sum_{p=O}^{2k-1}\frac{\gamma_p}{\delta_p}\right)\ln\frac{1}{r}-\lambda\sum_{p=O}^{2k-1}\frac{\gamma_p}{\delta_p}\ln\left(r-\frac{\delta_p}{c_p}R\right)$$

$$+\frac{\lambda}{2kc}\ln\frac{b+c\left(\frac{r}{R}\right)^{2k}}{b+c}-\frac{\lambda}{2k}\frac{a}{b}\ln\left(\frac{bR^{2k}+cr^{2k}}{b+c}\right).$$

But since

$$\frac{a}{b}=\frac{\bar{\lambda}}{P_{-k}},\qquad \frac{\gamma_p}{\delta_p}=\frac{P_{-k}^2-|\lambda|^2}{2k\lambda P_{-k}},\qquad \frac{\delta_p}{c_p}=\left(\frac{P_{-k}}{|\lambda|}\right)^{1/2k}e^{i(\varphi_p-\varphi)},$$

$$J=-P_{-k}\ln\frac{1}{r}+\ln\left(\frac{\lambda}{P_{-k}}+\frac{\bar{B}_k}{B_k}e^{-2ik\varphi}\right)+J_O(z), \tag{2.9}$$

where

$$J_O(z)=-\sum_{p=0}^{2k-1}\ln\left[\left(\frac{P_{-k}}{|\lambda|}\right)^{1/2k}e^{i\varphi_p}\left(\frac{c_p}{\delta_p}r-R\right)\right]$$

$$+\frac{P_{-k}}{2k}\ln\left[\frac{\bar{B}_k}{B_k}\left(1+\frac{c}{b}\left(\frac{r}{R}\right)^{2k}\right)\right]-\frac{|\lambda|^2}{2kP_{-k}}\ln\left[\frac{\bar{B}_k}{B_k}\left(R^{2k}+\frac{c}{b}r^{2k}\right)\right]$$

is a continuous function. Introducing (2.9) into (2.6), we obtain formula (2.3). Lemma 2.1 is proved.

2.2 Let us return to formula (2.1). Now it is possible to describe the properties of the analytic function $\Psi(z)$ at the point $z=O$. Substituting the expression of $\omega(z)$ from (2.3) we have

$$w(z)=\Psi(z)r^{P_{-k}}\left(1+\frac{P_{-k}}{\lambda}\frac{\bar{B}_k}{B_k}e^{-2ik\varphi}\right)e^{\omega_O^*(z)}. \tag{2.10}$$

This relation gives

2.1 Lemma 2.2 *The analytic function $\Psi(z)$ in* (2.1) *is continuous everywhere in G including the point $z = O$. Moreover*

$$\Psi(z) = O(|z|^k), \quad z \to 0, \tag{2.11}$$

k having the same value in formula (1.8).

In fact, as has been proved in Theorem 1.1, any continuous solution $w(z)$ of equation (1.1) certainly vanishes at the origin. Formula (1.8) indicates the possible zero orders, viz.

$$w(z) = O(r^{\mu_k}), \quad k = 0, 1, \ldots, \quad r \to 0.$$

Using this equality in the left side of formula (2.10) and dividing by r^{μ_k}, we get $\Psi(z)r^{-k} = O(1), r \to O$, and therefore (2.11). As to the continuity of $\Psi(z)$ outside the point $z = O$, this follows from (2.1) because of the continuity of $w(z)$ and $\omega(z)$.

Lemma 2.2 is proved.

Consequence 2.1 *If a continuous solution $w(z)$ of equation* (1.1) *has an infinite order of zero at the point $z = O$, then $w(z) \equiv O$ in G.*

In fact, from Lemma 2.2 one establishes that the analytic function $\Psi(z)$ has a zero of infinite order at the point $z = O$, but in this case $\Psi(z) \equiv O$ in G. Therefore we get $w(z) \equiv O$ from (2.1).

2.3 Now we consider the equation

$$\partial_{\bar{z}} w - \frac{b(z)}{2\bar{z}} \bar{w} = 0, \quad z \in E, \tag{2.12}$$

on the whole plane E, assuming that $b(z)$ in some domain G, containing the point $z = O$ inside, is subordinated to all the conditions listed in the introduction to the present chapter, and $b(z) \equiv O$ outside G.

Theorem 2.1 *Only $w(z) \equiv O$ is the unique solution of equation* (2.12), *continuous on the whole plane and vanishing at $z = \infty$.*

Proof Let us represent the solution of equation (2.12) in the form (2.1), where $\Psi(z)$ is an analytic function on the whole plane and $\omega(z)$ is determined by formula (2.2). Concerning continuous solutions $w(z)$ on the whole plane then $\omega(z)$ will also be continuous in $E - O$ and vanishes at infinity. The asymptotic behaviour of $\omega(z)$ in a neighbourhood of the point $z = O$ will be characterized by the relation (2.3). Under this condition $\Psi(z)$ in formula (2.1) is continuous in E and vaishes at infinity. Therefore $\Psi(z) \equiv O$ and so $w(z) \equiv O$ on account of (2.1).

Theorem 2.1 is proved.

3 The canonical form of the Riemann–Hilbert problem

Let G be a 1–connected domain of a complex plane z, bounded with Ljapunov contour Γ containing the origin.

Problem *It is necessary to find a function $w \in C(\bar{G}) \cap C_{\bar{z}}(G - O)$, satisfying in the domain G the equation*

$$\partial_{\bar{z}} w - \frac{b(z)}{2\bar{z}} \bar{w} = F(z) \tag{3.1}$$

and on the contour Γ the boundary problem

$$Re[g(z)w(z)] = h(z). \tag{3.2}$$

One assumes that $b(z), F(z), g(z), h(z)$ are given functions, $b(z)$ being continuous in $\bar{G}$ and satisfying the Holder condition with α and $b(O) \neq O; F(z) \in C(G - O) \cap L_p(\bar{G}), p > 2; g(z)$ and $h(z) \in C_\nu(\Gamma), 0 < \nu \leq 1, |g(z)| \neq 0$ on Γ.

This problem allows an essential simplifcation. First we show that without loss of generality the disk $|z| < R$ can be considered as the initial domain.

Indeed, according to the Riemann theorem a conformal mapping $\zeta = \zeta(z)$ of domain G onto some disk $|\zeta| < R$ exists so that $\zeta(O) = O$ and arg $\zeta(O) = 0$ [30].

Transforming equation (3.1) to a new variable, we get

$$\partial_{\bar{z}} w^* - \frac{b^*(z)}{2\bar{\zeta}} \overline{w^*} = F^*(\zeta), \quad |\zeta| \leq R,$$

in which

$$w^* = w(z(\zeta)), \quad F^*(z) = F(z(\zeta))\left(\frac{\overline{dz}}{d\zeta}\right); \quad b^*(\zeta) = \frac{\bar{\zeta}}{\bar{z}(\zeta)}\left(\frac{\overline{dz}}{d\zeta}\right) b(z(\zeta)).$$

It is easy to verify that $w^*(\zeta), b^*(\zeta)$ and $F^*(\zeta)$ in the disk $|\zeta| < R$ belong to the same classes of functions as $w(z), b(z), F(z)$ in the domain G. Moreover with regard to the mapping properties

$$z(\zeta) = \zeta + O(\zeta^2),$$

$$\frac{dz}{d\zeta} = 1 + O(\zeta), \quad \zeta \to 0,$$

we have

$$b^*(0) = b(0).$$

Therefore in the conformal mapping of the domain G in the disk not only the form of the initial equation remains but also the value of $b(z)$ at the singular point.

Now we modify the boundary condition (3.2). Let $m = \mathrm{Ind}_\Gamma g(z)$ be the index of $g(z)$ along the contour Γ. According to [23] for $g(z)$, there exists a real regularizing multiplier $p(z), z \in \Gamma$, such that

$$p(z)g(z) = z^m e^{i\gamma(z)}, \quad z \in \Gamma,$$

where $\gamma(z)$ is the boundary value of the function analytic in G and not vanishing in this domain.

Using $p(z)$, we can rewrite equality (3.2) in the form

$$Re\left[z^m e^{i\gamma(z)} w(z)\right] = h^*(z),$$

where $h^*(z) = p(z)h(z)$.

We further introduce the new unknown function

$$w^*(z) = e^{i\gamma(z)} w(z), \quad z \in G,$$

in which equation (3.1) takes the form:

$$\partial_{\bar z} w^* - \frac{b^*(z)}{2\bar z}\overline{w^*} = F^*(z).$$

Here $b^*(z) = b(z)e^{-i(\bar\gamma+\gamma)}$ and $F^*(z) = F(z)e^{-i\gamma}$. Moreover $w^*(z), z \in \Gamma$, is subordinated to the condition

$$Re\left[z^m w^*(z)\right] = h^*(z),$$

Thus, we could transform the boundary condition (3.2) into a specific form, preserving the original form of the equation. We should note that when transferring to a new unknown function the value of the modulus of the coefficient $b(z)$ remains invariant in the singular point, i.e. $|b(0)| = |b^*(0)|$.

So without loss of generality boundary problem (3.1), (3.2) is reduced to the following canonical form:

$$\partial_{\bar{z}} w - \frac{b(z)}{2\bar{z}}\bar{w} = F(z), \quad z \in G, \tag{3.3}$$

$$Re\left[z^{-m}w(z)\right] = h(z), \quad z \in \Gamma, \tag{3.4}$$

where $G = \{z : |z| < R\}$ and $\Gamma = \{z : z = Re^{i\varphi}\}$.

4 The Riemann–Hilbert problem with a zero index [55]

In the present section we restrict ourselves to the case

$$Re\, w = h(z), \quad |z| = R. \tag{4.1}$$

We consider this case separately because of its specific character and the fact that its results do not arise from the case $m > 0$ or $m < 0$.

4.1 Let us derive a necessary condition for the solvability of problem (3.3), (4.1) with $b(0) < 0$. To this end we write the equation in the following way:

$$\partial_{\bar{z}} w - \frac{b(0)}{2\bar{z}}\bar{w} = f(z, w), \tag{4.2}$$

where

$$f(z, w) = F(z) + \frac{b(z) - b(0)}{2\bar{z}}\overline{w(z)}.$$

If we assume that f is a known function, then the homogeneous boundary problem, conjugate to (4.2), (4.1), will be the following:

$$\partial_{\bar{z}} U + \frac{b(0)}{2\bar{z}} U = 0, \quad |z| < R, \tag{4.3}$$

$$\mathrm{Re}(ie^{i\varphi}U) = 0 \quad \text{on} \quad |z| = R, \tag{4.4}$$

where $\varphi = \arg z$.

In the class of functions continuous outside $z = O$ and admitting no more than a first-order singularity at $z = O$, problem (4.3), (4.4) has only the non-trivial solution

$$U(z) = \frac{1}{z}|z|^{-b(0)}. \tag{4.5}$$

Now we use Green's identity (see [69], §9, ch. 3) in two arbitrary solutions of equations (4.2) and (4.3):

$$\mathrm{Re}[\frac{1}{2\mathrm{i}}\int_{\Gamma+\Gamma_\varepsilon} wU\,\mathrm{d}z] = \mathrm{Re}\iint\limits_{G-G_\varepsilon} Uf\,\mathrm{d}x\,\mathrm{d}x, \quad z = x+\mathrm{i}y,$$

where G_ε is the disk $|z| \le \varepsilon$ and Γ_ε is the circle $|z| = \varepsilon$. If in this identity in place of U we take the right side of equality (4.5), and instead of f we employ its expression, then after tending to the limit as $\varepsilon \to O$ we get

$$T_G\left(\frac{b(z)-b(0)}{|z|}\overline{w(z)}\right) = h_O - T_G\big(2\mathrm{e}^{-\mathrm{i}\varphi}F(z)\big). \tag{4.6}$$

Here

$$h_O = \frac{1}{2\pi}\int_O^{2\pi} h(\theta)\,\mathrm{d}\,\theta \quad \text{and} \quad T_G f = \frac{R^{b(0)}}{2\pi}\mathrm{Re}\iint\limits_G |z|^{|b(0)|-1} f(z)\,dx\,dy$$

is a linear bounded functional in $L_p, p > 2$.

Thus, if $w(z)$ is a solution of the boundary problem (3.3), (4.1) with $b(0) < O$, then it must satisfy relation (4.6).

4.2 We define a pair of functions Ω_1^O and Ω_2^O in the domain $|z|, |\zeta| \le R$, by the equalities

$$\begin{aligned}\Omega_1^O =& \frac{\delta(|\lambda|-\bar\lambda)}{4|\lambda|}\left(\frac{r\rho}{R^2}\right)^{|\lambda|} \\ &+ \sum_{k=1}^{\infty}\left[\frac{\lambda(P_{-k}-\bar\lambda)}{2\mu_k(P_{-k}+\lambda)}\mathrm{e}^{\mathrm{i}k(\varphi-\gamma)} + \frac{P_{-k}(P_k-\bar\lambda)}{2\mu_k(P_{-k}+\bar\lambda)}\mathrm{e}^{-\mathrm{i}k(\varphi-\gamma)}\right]\left(\frac{r\rho}{R^2}\right)^{\mu_k}, \\ \Omega_2^O =& \frac{\delta(|\lambda|-\bar\lambda)}{4|\lambda|}\left(\frac{r\rho}{R^2}\right)^{|\lambda|} \\ &+ \sum_{k=1}^{\infty}\left[\frac{\lambda(P_k-\lambda)}{2\mu_k(P_{-k}+\lambda)}\mathrm{e}^{\mathrm{i}k(\varphi-\gamma)} + \frac{P_{-k}(P_{-k}-\lambda)}{2\mu_k(P_{-k}+\bar\lambda)}\mathrm{e}^{-\mathrm{i}k(\varphi-\gamma)}\right]\left(\frac{r\rho}{R^2}\right)^{\mu_k}.\end{aligned}$$

Here

$$\delta = \begin{cases} 1 & \text{for } \lambda > O \\ 0 & \text{for } \lambda < O \\ \mathrm{i}(|\lambda|-\lambda)/\mathrm{Im}\,\lambda & \text{for } \mathrm{Im}\,\lambda \ne O. \end{cases}$$

The meaning of other notation has been given in §1, ch. 1.

One establishes by immediate verification that Ω_1^O and Ω_2^O are continuous for $|z|, |\zeta| < R$, continuously differentiable with respect of z and $\bar{z}$ outside $z = O$ and satisfy the system

$$\partial_{\bar{z}}\Omega_1^O - \frac{\lambda}{2\bar{z}}\overline{\Omega_2^O} = O, \qquad \partial_{\bar{z}}\Omega_2^O - \frac{\lambda}{2\bar{z}}\overline{\Omega_1^O} = O. \tag{4.7}$$

For Ω_1^O the inequality

$$\sup_{|z|\leq R}\left(\iint_{|\zeta|\leq R} |\Omega_1^O|^p \, d\xi \, d\eta\right)^{1/p} < K_1 R^{2/p}, \tag{4.8}$$

is valid where p is any number greater than 1 and K_1 is a constant depending on p and R. The same inequality but with constant K_2 holds for

$$\Omega_2^O - R^2/(R^2 - z\bar{\zeta}).$$

The pair of functions Ω_1^O and Ω_2^O together with Ω_1 and Ω_2, which are defined in §1, ch. 1, satisfy the relation

$$\left[\Omega_1 + \Omega_1^O + \bar{\Omega}_2 + \overline{\Omega_2^O}\right]_{|z|=R} = \begin{cases} -(\rho/R)^{|\lambda|} & \text{for } \lambda < O, \\ 0 & \text{for other } \lambda. \end{cases}$$

4.3 Now we examine the integral operator

$$S_G^O f = -\frac{1}{\pi}\iint_G \left[\frac{\Omega_1^O(z,\zeta)}{\zeta} f(\zeta) + \frac{\Omega_2^O(z,\zeta)}{\bar{\zeta}}\overline{f(\zeta)}\right] d\xi d\eta,$$

where the domain G is the disk $|z| \leq R$, and Ω_1^O, Ω_2^O are given in subsection 4.2. On the basis of the properties of the functions Ω_1^O, Ω_2^O it is not hard to state the following proposition.

Lemma 4.1 *The operator S_G^O is completely continuous from $L_p(\bar{G}), q < 2$, in $C(\bar{G})$. In addition*

$$\|S_G^O\|_C \leq N^O R^{\frac{2-p}{p}},$$

where N^O is a constant independent of R, and $p^{-1} + q^{-1} = 1$.

The function $S_G^O f$ is a continuous solution of the equation

$$\partial_{\bar{z}}\Phi - \frac{\lambda}{2\bar{z}}\bar{\Phi} = 0, \quad z \in G. \tag{4.9}$$

Together with the function $S_G f$, given by formula (1.11), ch.1, it satisfies the relation:

$$Re\big(S_G f + S_G^O\big)_{|z|=R} = \begin{cases} T_G(2e^{-i\gamma} f(\zeta)) & \text{for } \lambda < O, \\ 0 & \text{for other } \lambda. \end{cases} \tag{4.10}$$

The operator S_G^O serves as a basis for constructing another important operator:

$$P_G^O \bar{w} = S_G^O\left(\frac{b(\zeta) - b(O)}{2\bar{\zeta}}\overline{w(\zeta)}\right).$$

Its properties are characterized by

Lemma 4.2 *The operator P_G^O is completely continuous in $C(\bar{G})$,* mapping this space into itself. Moreover

$$\|P_G^O\|_C < N^O R^{\frac{2-p}{p}} \left\|\frac{b(z) - b(O)}{2\bar{z}}\right\|_{L_q}, \quad q > 2, \tag{4.11}$$

and the function $P_G^O \bar{w}$ is a continuous solution of equation (4.9). *Jointly with $P_G \bar{w}$, defined by formula* (4.3), ch.1 *it satisfies the relation*

$$\mathrm{Re}\big(P_G \bar{w} + P_G^O \bar{w}\big)_{|z|=R} = \begin{cases} T_G\left(\frac{b(\zeta)-b(O)}{|\zeta|}\overline{w(\zeta)}\right) & \text{for } \lambda < O, \\ 0 & \text{for other } \lambda \end{cases}. \tag{4.12}$$

4.4 We proceed to the solution of boundary problem (3.3), (4.1). We keep in mind that $\lambda = b(O)$. With regard to Theorem 4.1, ch. 1, the continuous solution $w(z)$ of equation (3.3) satisfies integral equation (4.2) from §4, ch. 1. A solution of the latter is not obliged to satisfy boundary condition (4.1). However, on the basis of this equation a new integral equation is constructed which is equivalent to boundary problem (3.3), (4.1). For this purpose we represent a solution $\Phi(z)$ of model equation (4.9) in the form

$$\Phi(z) = \Phi_O(z) + S_G^O F + P_G^O \bar{w},$$

where S_G^O and P_G^O are operators, defined in subsection 4.3, and $\Phi_O(z)$ is a new unknown solution of the model equation, which is continuous in $\bar{G}$.

Introducing the expression for $\Phi(z)$ into (4.2) from §4, ch.1 we get

$$w(z) - \hat{P}_G \bar{w} = \Phi_O(z) + \hat{S}_G F, \tag{4.13}$$

where $\hat{P}_G = P_G + P_G^O, \hat{S}_G = S_G + S_G^O, \hat{P}_G$ being completely continuous in $C(\bar{G})$ and by virtue of (4.11) above and (4.4) from ch.1 satisfying the inequality

$$\|\hat{P}_G\|_C < \hat{N} R^{\mathcal{X}_O} \left\| \frac{b(z) - b(O)}{2\bar{z}} \right\|_{L_q}, \quad q > 2. \tag{4.14}$$

Here $\hat{N}$ is a constant independent of R, and $\mathcal{X}_O > O$.

If we assume $|z| = R$ in (4.13) and take the real part, then by virtue of boundary condition (4.1) we have

$$h(z) - \text{Re } \left[\hat{P}_G \bar{w}\right]_\Gamma = \text{Re } \left[\Phi_O(z)\right]_\Gamma + \text{Re}\left[\hat{S}_G F\right]_\Gamma. \tag{4.15}$$

Let $b(O) > O$ or Im $b(0) \neq 0$. Then with regard of (4.10), (4.12) from (4.15) we get

$$\text{Re}\left[\Phi_O(z)\right]_\Gamma = h(z). \tag{4.16}$$

Thus $\Phi_O(z)$ is a continuous solution of boundary problem (4.9), (4.16) which according to Theorem 9.1, ch. 2, has a unique solution. Introducing this solution into (4.13) we obtain an integral solution of Fredholm type which is equivalent to boundary problem (3.3), (4.1). For $\|\hat{P}_G\|_C < 1$ this problem will obviously have a unique solution.

Now let $b(0) < O$. According (4.10), (4.12) from (4.13) we get

$$\text{Re}\left[\Phi_O(z)\right]_\Gamma = h(z) - T_G\left(\frac{b(\zeta) - b(O)}{|\zeta|}\overline{w(\zeta)}\right) - T_G\left(2\text{e}^{-\text{i}\gamma} F(\zeta)\right).$$

Transforming the right side with regard to (4.6), we come to the boundary condition

$$\text{Re}\left[\Phi_O(z)\right]_\Gamma = h(z) - h_O. \tag{4.17}$$

From Theorem 9.1, ch. 2, it follows that the solution of problem (4.9)(4.17) is given by the formula

$$\Phi_O(z) = \text{i}cr^{-b(O)} + H(z).$$

In connection with this, (4.13) takes the form

$$w(z) - \hat{P}_G \bar{w} = \text{i}cr^{-b(O)} + H(z) + \hat{S}_G F. \tag{4.18}$$

Thus, boundary problem (3.3), (4.1) is equivalent to integral equation (4.18) with the additional condition (4.6).

Let $\|\hat{P}_G\|_C < 1$. The solution to equation (4.18) can obviously be expressed in the form $w = w_O + cw_1$, where

$$w_O - \hat{P}_G\overline{w_O} = H(z) + \hat{S}_G F$$
$$w_1 - \hat{P}_G\overline{w_1} = \mathrm{i}r^{-b(O)}.$$

Substituting in (4.6) we get

$$cT_G\left(\frac{b(z)-b(O)}{|z|}\overline{w_1(z)}\right) = h_O - T_G\big(2\mathrm{e}^{-\mathrm{i}\varphi}F(z) + \frac{b(z)-b(O)}{|z|}\overline{w_O(z)}\big).$$

Is it possible to fulfil this relation having chosen c. This depends on the value

$$\tau = T_G\left(\frac{b(z)-b(O)}{|z|}\overline{w_1(z)}\right).$$

If $\tau \neq 0$, then c is uniquely determined and problem (3.3), (4.1) will have a unique solution. But if $\tau = 0$, then the homogeneous problem (3.3), (4.1) ($F = h = 0$) will have only one non–trivial solution, and the inhomogeneous problem will be solvable under the condition

$$T_G\left(2e^{-\mathrm{i}\varphi}F(z) + \frac{b(z)-b(O)}{|z|}\overline{w_O(z)}\right) = h_O.$$

Since the value τ is determined by a finite calculation from $b(z)$ the set of all admissible coefficients should be split into two disjoint subsets B_1 and B_2 so that $b(z) \in B_1$, if $\tau = 0$, and $b(z) \in B_2$ if $\tau \neq 0$. These subsets are not empty. Indeed, $b(z) \equiv b(0) < 0$ belongs to B_1. The solvability condition takes the form

$$T_G(2e^{-\mathrm{i}\varphi}F(z)) = h_O.$$

One of the elements of B_2 is $b(z) = -1 + ir$. Let us summarize this discussion.

Theorem 4.1 *Let the inequality* $\|\hat{P}_G\|_C < 1$ *be satisfied. Then for* $b(0) > 0$ *or* Im $b(0) \neq 0$, *also for* $b(0) < O$ *with* $b(z) \in B_2$, *the solution of* (3.3), (4.1) *exists and is unique.*

For $b(0) < O$ *with* $b(z) \in B_1$ *the homogeneous problem* ($F = h = 0$) *has exactly one non–trivial solution, and for the existence of a solution of the inhomogeneous problem it is necessary and sufficient to meet the requirement of the solvability condition on* F *and* h.

4.5 If the contractness condition does not hold, then from the complete continuity of the operator $\hat{P}_G$ a more general assertion is valid.

Theorem 4.1′ *The index* $\mathcal{X}$ *of problem* (3.3), (4.1) *is equal to zero.*

Here $\mathcal{X} = l - l'$, where l is the number of linearly independent solutions of the homogeneous boundary problem and l' is the number of real solvability conditions of the inhomogeneous problem.

5 The Riemann–Hilbert problem with a positive index

We study the problem of finding an unknown function $w(z)$ satisfying the equation

$$\partial_{\bar{z}} w - \frac{b(z)}{2\bar{z}} \bar{w} = F(z), \quad |z| < R, \tag{5.1}$$

and the boundary condition

$$\operatorname{Re}\left[z^{-m} w\right] = h(z) \quad \text{on } |z| = R. \tag{5.2}$$

One assumes that $m > O$. As in §4, the principal point of the investigation is the construction of an additional integral operator by which the present problem is reduced to an analogous one for the model equation.

5.1 In the set $|z|, |\zeta| \leq R$, we introduce the functions $\Omega_1^+(z,\zeta,m)$ and $\Omega_2^+(z,\zeta,m)$ by the equalities

$$\begin{aligned}
\Omega_1^+(z,\zeta,m) =& \frac{1}{2}\left[A_O(z) + C_O(z)\right]\left[N_m(\zeta) + M_m(\zeta)\right] \\
&+ \sum_{p=1}^{\infty} A_p(z)\left[N_{m+p}(\zeta) + \overline{M_{m-p}(\zeta)}\right] + C_p(z)\left[\overline{M_{m+p}(\zeta)} + N_{m-p}(\zeta)\right], \\
\Omega_2^+(z,\zeta,m) =& \frac{1}{2}\left[A_O(z) + C_O(z)\right]\left[\overline{N_m(\zeta)} + M_m(\zeta)\right] \\
&+ \sum_{p=1}^{\infty} A_p(z)\left[M_{m+p}(\zeta) + \overline{N_{m-p}(\zeta)}\right] + C_p(z)\left[\overline{N_{m+p}(\zeta)} + M_{m-p}(\zeta)\right].
\end{aligned}$$

In these formulas we use the notation:

$$A_p(z) = e^{i(m+p)\varphi}\left(\frac{r}{R}\right)^{\mu_{m+p}} + \sum_{k=1}^{\infty}\frac{(-1)^k}{\lambda^k}\left(\prod_{\delta=1}^{k}P_{-(2\delta-1)m-p}\right)$$

$$\times\, e^{i[(2k+1)m+p]\varphi}\left(\frac{r}{R}\right)^{\mu_{(2k+1)m+p}};$$

$$C_p(z) = \sum_{k=O}^{\infty}\frac{(-1)^k}{\bar\lambda^{k+1}}\left(\prod_{\delta=1}^{k+1}P_{-(2\delta-1)m-p}\right)e^{-i[(2k+1)m+p]\varphi}\left(\frac{r}{R}\right)^{\mu_{(2k+1)m+p}};$$

$$N_k(\zeta) = \frac{P_{-k}}{2\mu_k}e^{-ik\gamma}\left(\frac{\rho}{R}\right)^{\mu_k}; \qquad M_k(\zeta) = -\frac{\lambda}{2\mu_k}e^{-ik\gamma}\left(\frac{\rho}{R}\right)^{\mu_k}; \qquad \lambda = b(0).$$

We recall that $z = re^{i\varphi}, \zeta = \rho e^{i\gamma}, \mu_k^2 = |\lambda|^2 + k^2, P_{-k} = \mu_k - k$ and $\prod$ is the product symbol.

The functions Ω_1^+ and Ω_2^+ have the following properties. They are continuous for $|z|, |\zeta| < R$, continuously differentiable with respect to z and $\bar z$ outside $z = O$ and satisfy the system of equations (4.7). The functions Ω_1^+ and $\Omega_2^+ - R^2 e^{im(\varphi+\gamma)}$ are subject to inequality (4.8). Together with Ω_1 and Ω_2 (see §1, ch. 1) they satisfy the relation

$$z^{-m}(\Omega_1 + \Omega_1^+) + \overline{z^{-m}(\Omega_2 + \Omega_2^+)} = 0 \quad \text{on } |z| = R.$$

On the basis of $\Omega_1^+(z,\zeta,m)$ and $\Omega_2^+(z,\zeta,m)$ we construct the operators

$$S_G^+ f = -\frac{1}{\pi}\iint_G\left[\frac{\Omega_1^+(z,\zeta,m)}{\zeta}f(\zeta) + \frac{\Omega_2^+(z,\zeta,m)}{\bar\zeta}\overline{f(\zeta)}\right]d\xi\, d\eta,$$

$$P_G^+\bar w = S_G^+\left(\frac{b(\zeta)-b(0)}{2\bar\zeta}\overline{w(\zeta)}\right)$$

with the properties enumerated in Lemmas 4.1 and 4.2 for the operators S_G^O and P_G^O, only instead (4.10) and (4.12) the following relations are valid

$$\mathrm{Re}\left[z^{-m}(S_G f + S_G^+ f)\right] = 0,$$

$$\mathrm{Re}\left[z^{-m}(P_G\bar w + P_G^+\bar w)\right] = 0 \quad \text{on } |z| = R.$$

5.2 Now we solve the problem (5.1), (5.2) with $m > O$. On the basis of (4.4), ch.1 we derive an integral equation equivalent to boundary problem (5.1), (5.2). To this end we represent

$$\Phi(z) = \Phi_O(z) + S_G^+ F + P_G^+\bar w,$$

where S_G^+ and P_G^+ are the operators of subsection 5.1 and $\Phi_0(z)$ is a new unknown solution of model equation (4.9). Substituting $\Phi(z)$ in (4.4), ch.1, we get

$$w(z) - P_G^* \bar{w} = \Phi_O(z) + S_G^* F, \tag{5.3}$$

where $P_G^* = P_G + P_G^+$ and $S_G^* = S_G + S_G^+$, the operator P_G^* being completely continuous in $C(G)$ and subordinating to inequality (4.14).

Introducing $w(z)$ from (5.3) into boundary condition (5.2) and taking into account the above properties of $S_G^+ f$ and $P_G^+ \bar{w}$ on $|z| = R$ we get

$$\mathrm{Re}(z^{-m}\Phi_O(z)) = h(z) \quad \text{on } \Gamma. \tag{5.4}$$

Thus, any continuous solution of integral equation (5.3) will automatically satisfy boundary condition (5.2) if $\Phi_O(z)$ is a solution of the boundary problem (4.9), (5.4). The general solution of this problem was described in Theorem 9.2, ch. 2. Introducing this solution for $\Phi_O(z)$ into (5.3) we get

$$w(z) - P_G^* \bar{w} = c_O \tilde{\Phi}_O(z) + \sum_{k=1}^{m-1} c_k^{(1)} \tilde{\Phi}_k(z) + c_k^{(2)} \tilde{\tilde{\Phi}}_k(z) + c_m \tilde{\Phi}_m(z) + H(z) + S_G^* F. \tag{5.5}$$

Theorem 5.1 *For $m > O$ the boundary problem* (5.1), (5.2) *in the class of continuous functions is equivalent to the integral equation* (5.3).

Under the condition $\|P_G^*\|_C < 1$ the homogeneous problem $(F = h = 0)$ has $2m$ linearly independent solutions, and the inhomogeneous problem is solvable for any $h(z)$ and $F(z)$.

The assertions of this theorem are evident. If the contractness condition of operator P_G^* is not valid, as it was earlier, then we should employ a more general asssertion.

Theorem 5.1 *The index $\mathcal{X}$ of problem* (5.1), (5.2) *is equal to $2m$.*

6. The Riemann–Hilbert problem with a negative index

6.1 We deduce necessary conditions for the solvability of problem (5.1), (5.2) with $m < 0$. To do this we write equation (5.1) in the form (4.2). If one assumes that f is a known function then the homogeneous boundary problem which is conjugate to (4.2), (5.2), will be the following, $(\lambda = b(0))$ according to [69]:

$$\partial_{\bar{z}} U + \frac{\bar{\lambda}}{2\bar{z}} \bar{U} = 0, \quad |z| < R, \tag{6.1}$$

$$\operatorname{Re}\left[\mathrm{i}e^{\mathrm{i}(m+1)\varphi}U\right] = 0 \quad \text{on } |z| = R. \tag{6.2}$$

Let us write out Green's identity ([69], §2, ch. 4) for two arbitrary solutions of the above boundary problems:

$$\operatorname{Re} \iint\limits_{G-G_\varepsilon} Uf \,\mathrm{d}x\,\mathrm{d}y = \frac{1}{2\mathrm{i}} \int_{\Gamma+\Gamma_\varepsilon} z^m U h \,\mathrm{d}z,$$

where G and G_ε are the domains $|z| \le R$ and $|z| \le \varepsilon$ respectively, while Γ and Γ_ε are the circles $|z| = R$ and $|z| = \varepsilon$. If in this identity we take as U one of the linearly independent solutions of problem (6.1), (6.2) and use the notation given for $f(z,w)$ then passing to the limit as $\varepsilon \to 0$ we obtain

$$T_p\left(\frac{b(z)-b(0)}{2\bar{z}}\overline{w(z)}\right) + T_p(F(z)) = B_p, \quad p = 0, 1, \ldots, |m|. \tag{6.3}$$

Here for $p = 0$ and $p = |m|$

$$T_p(q) = \frac{R^{-m-1}}{\pi} \iint\limits_G \operatorname{Re}\left[U_p(z)q(z)\right] \mathrm{d}x\,\mathrm{d}y,$$

and for the values $p = 1, \ldots, |m| - 1$

$$T_p(q) = \frac{R^{-m-1}}{\pi} \iint\limits_G \{\operatorname{Re}[U_p(z)q(z)] + \mathrm{i}\operatorname{Re}[U_p^*(z)q(z)]\} \,\mathrm{d}x\,\mathrm{d}y.$$

We recall that the constant B_p was defined in subsection 9.4, ch.2.

Thus, if $w(z)$ is any continuous solution of boundary problem (5.1), (5.2) with $m < 0$, then it must satisfy conditions (6.3).

6.2 As in §§4 and 5 we introduce the functions $\Omega_1^-(z,\zeta,m)$ and $\Omega_2^-(z,\zeta,m)$ which will be used in future as the kernels of an auxiliary integral operator. The expressions for these functions are bulky:

$$\begin{aligned}
\Omega_1^-(z,\zeta,m) = & \frac{1}{2}\left(\frac{r}{R}\right)^{\mu_O} \left[B_{O,-|m|}(\zeta) + \frac{\mathcal{X}}{\bar{\mathcal{X}}}\overline{C_{O,-|m|}(\zeta)}\right] \\
& + \sum_{p=0}^{|m|-1} \sum_{k=0}^{\infty} \left[A_{k,-p}(z)\left(\frac{\lambda P_{-[(2k+1)|m|-p]}}{2\mu_{[(2k+1)|m|-p]}}\overline{A_{k,-p}(\zeta)} + B_{k,-p}(\zeta)\right)\right. \\
& \left. + \overline{A_{k,-p}(z)} P_{-[(2k+1)|m|-p]}\left(-\frac{\bar{\lambda}}{2\mu_{[(2k+1)|m|-p]}} A_{k,-p}(\zeta) + \overline{C_{k,-p}(\zeta)}\right)\right]
\end{aligned}$$

$$+\sum_{p=1}^{|m|}\sum_{k=0}^{\infty}\Bigg[A_{k,p}(z)\Big(\frac{\lambda P_{-[(2k+1)|m|+p]}}{2\mu_{[(2k+1)|m|+p]}}\overline{A_{k,p}}(\zeta)+B_{k,p}(\zeta)\Big)$$

$$+\overline{A_{k,p}}(z)P_{-[(2k+1)|m|+p]}\Big(-\frac{\bar\lambda}{2\mu_{[(2k+1)|m|+p]}}\overline{A_{k,p}}(\zeta)+C_{k,p}(\zeta)\Big)\Bigg];$$

$$\Omega_2^-(z,\zeta,m)=\frac{1}{2}\Big(\frac{r}{R}\Big)^{\mu_O}\Big[C_{O,-|m|}(\zeta)+\frac{\mathcal{X}}{\bar{\mathcal{X}}}\overline{B_{O,-|m|}}(\zeta)\Big]$$

$$+\sum_{p=0}^{|m|-1}\sum_{k=0}^{\infty}\Bigg[A_{k,-p}(z)\Big(-\frac{\lambda^2}{2\mu_{[(2k+1)|m|-p]}}\overline{A_{k,-p}}(\zeta)+C_{k,-p}(\zeta)\Big)$$

$$+\overline{A_{k,-p}}(z)P_{-[(2k+1)|m|-p]}\Big(\frac{P_{-[(2k+1)|m|-p]}}{2\mu_{[(2k+1)|m|-p]}}A_{k,-p}(\zeta)+\overline{B_{k,-p}}(\zeta)\Big)\Bigg]$$

$$+\sum_{p=1}^{|m|}\sum_{k=0}^{\infty}\Bigg[A_{k,p}(z)\Big(-\frac{\lambda^2}{2\mu_{[(2k+1)|m|+p]}}\overline{A_{k,p}}(\zeta)+C_{k,p}(\zeta)\Big)$$

$$+\overline{A_{k,p}}(z)P_{-[(2k+1)|m|+p]}\Big(\frac{P_{-[(2k+1)|m|+p]}}{2\mu_{[(2k+1)|m|+p]}}A_{k,p}(\zeta)+\overline{B_{k,p}}(\zeta)\Big)\Bigg].$$

Here we have applied the notation

$$A_{s,\pm p}(z)=e^{i[(2s+1)|m|\pm p]\varphi}\Big(\frac{r}{R}\Big)^{\mu_{[(2(s+1)|m|\pm p]}},$$

$$B_{s,\pm p}(\zeta)=\sum_{\beta=1}^{\infty}\frac{(-1)^\beta}{\lambda^\beta}\prod_{\delta=1}^{\beta}P_{-[(2\delta+2s+1)|m|\pm p]}\overline{A_{s+\beta,\pm p}}(\zeta);$$

$$C_{s,\pm p}(\zeta)=\sum_{\beta=0}^{\infty}\frac{(-1)^\beta}{\lambda^\beta}\prod_{\delta=1}^{\beta}P_{-[(2\delta+2s+1)|m|\pm p]}\overline{A_{s+\beta+1,\pm p}}(\zeta).$$

The functions Ω_1^- and Ω_2^- are continuous for $|z|,|\zeta|<R$, continuously differentiable with respect to z and $\bar z$ outside $z=0$ and satisfy (4.7). Both functions are subordinated to inequality (4.8). Together with the functions Ω_1 and Ω_2 they satisfy the relation

$$\Big[z^{-m}(\Omega_1+\Omega_1^-)+\overline{z^{-m}(\Omega_2+\Omega_2^-)}\Big]_{|z|=R} \tag{6.4}$$

$$=\zeta R^{-m-1}\big[U_O(\zeta)\nu_O(\varphi)+U_{|m|}(\zeta)\nu_{|m|}(\varphi)+\sum_{k=1}^{|m|-1}U_k(\zeta)\cos k\varphi-U_k^*(\zeta)\sin k\varphi\big],$$

where the meaning of the notation has been indicated in subsection 9.4, ch.2. On the basis of Ω_1^- and Ω_2^-, the following operators are constructed:

$$S_G^- f = -\frac{1}{\pi}\iint\limits_G \left[\frac{\Omega_1^-(z,\zeta,m)}{\zeta} f(\zeta) + \frac{\Omega_2^-(z,\zeta,m)}{\bar{\zeta}}\overline{f(\zeta)}\right] d\xi\, d\eta,$$

$$P_G^- \bar{w} = S_G^-\left(\frac{b(\zeta)-b(0)}{2\bar{\zeta}}\overline{w(\zeta)}\right).$$

Their properties coincide with those enumerated for S_G^O and P_G^O in Lemmas 4.1 and 4.2.

6.3 We proceed directly to investigating boundary problem (5.1), (5.2) with $m < 0$. We pass over to integral equation (4.2) from ch. 1 and represent $\Phi(z)$ in the form

$$\Phi(z) = \Phi_O(z) + S_G^- F + P_G^- \bar{w},$$

where S_G^- and P_G^- are the above operators and $\Phi_O(z)$ is a new unknown solution of equation (4.9). Substituting $\Phi(z)$ into (4.2) we get

$$w(z) - P_G^{**}\bar{w} = \Phi_O(z) + S_G^{**}F, \tag{6.5}$$

where $P_G^{**} = P_G + P_G^-$ and $S_G^{**} = S_G + S_G^-$. The operator P_G^{**} is completely continuous in $C(G)$ and maps this space to itself. In addition P_G^{**} is subordinated to an inequality of type (4.11).

Introducing $w(z)$ from (6.5) into boundary condition (5.2) and taking (6.4) into account we have the relation

$$\begin{aligned}\operatorname{Re}(z^{-m}\Phi_O)_\Gamma &- \left\{T_O(F(z)) + T_O\left(\frac{b(z)-b(O)}{2\bar{z}}\overline{w(z)}\right)\right\}v_O(\varphi)\\ &- \left\{T_{|m|}(F(z)) + T_{|m|}\left(\frac{b(z)-b(0)}{2\bar{z}}\overline{w(z)}\right)v_{|m|}(\varphi)\right.\\ &- \sum_{k=1}^{|m|-1} \operatorname{Re}\left\{T_k\Big(F(z)\Big) + T_k\Big(\frac{b(z)-b(O)}{2\bar{z}}\overline{w(z)}\Big)\right\}\cos k\varphi\\ &+ \sum_{k=1}^{|m|-1} \operatorname{Im}\left\{T_k(F(z)) + T_k\left(\frac{b(z)-b(0)}{2\bar{z}}\overline{w(z)}\right)\right\}\sin k\varphi = h(z).\end{aligned}$$

This expression with regard to the solvability condition (6.3) is transformed into boundary condition (9.41), ch.2, for the unknown function $\Phi_O(z)$.

According to Consequence 9.1, ch. 2, the above problem has a unique solution. Denoting it by $H_O(z)$ and substituting it instead of $\Phi_O(z)$ into (6.5) we get

$$w(z) - P_G^{**}\bar{w} = H_O(z) + S_G^{**}F. \tag{6.6}$$

Theorem 6.1 *For $m < O$ the boundary problem* (5.1),(5.2) *equivalent to the integral equation* (6.6) *with supplementary condition* (6.3).

The homogeneous problem ($F = h = 0$) has only the zero solution. For the existence of a solution of the inhomogeneous problem it is necessary (and sufficient under the assumption $\|P_G^{**}\|_C < 1$) that $h(z)$ and $F(z)$ satisfy $2|m|$ real relations.

Let us analyse the homogeneous problem. Substituting the right side of formula (2.10) into (5.2) instead of $w(z)$ we get the Riemann–Hilbert homogeneous problem for the analytic function $\Psi(z)$. Since the index of the problem is negative ($m < 0$), then $\Psi(z) \equiv 0$ [23, 39]. In this case from (2.10) it follows that $w(z) \equiv 0$, i.e. the homogeneous problem (5.1), (5.2) has only the zero solution.

We consider further the index $\mathcal{X} = l - l'$ of boundary problem (5.1), (5.2), where l is the number of linearly independent solutions of the homogeneous problem (in this case $l = 0$) and l' is the number of real solvability conditions of the inhomogeneous problem. The problem index is easy to compute: $\mathcal{X} = 2m$. Hence $l' = 2|m|$.

The other assertions of the theorem are evident.

7 The conjugation problem

We denote $G^+ = \{z : |z| < R\}, \Gamma = \{z : |z| = \mathrm{Re}^{i\varphi}\}$ and $G^- = \{z : |z| > R\}$. Let $w^+(z)$ be a function of the class $C(G^+) \cap C_{\bar{z}}(G^+ - O)$, satisfying the equation

$$\partial_{\bar{z}} w^+ - \frac{b(z)}{2\bar{z}}\overline{w^+} = F(z), \quad z \in G^+. \tag{7.1}$$

We accept the same assumptions of the properties of $b(z)$ and $F(z)$ as in §3.

We consider an analytic function $\Phi^-(z)$ in G^- bounded at $z = \infty$ and continuously extendable on Γ.

Problem *It is necessary to find functions $w^+(z), z \in G^+$ and $\Phi^-(z), z \in G^-$, which are connected by the relation*

$$w^+(z) = z^m \Phi^-(z) + g(z), \quad z \in \Gamma, \tag{7.2}$$

where $g(z)$ is a given function satisfying the Holder condition.

Just as before we use an integral equation on a domain for investigating the above problem.

7.1 Let us define in the domain $|z|, |\zeta| < R$ the pair of functions $\Omega_1^+(z,\zeta)$ and $\Omega_2^+(z,\zeta)$ by the equalities

$$\Omega_1^+ = \sum_{k=1}^{\infty} \frac{P_{-k}}{2\mu_k}\left[e^{ik(\varphi-\gamma)} - e^{-ik(\varphi-\gamma)}\right]\left(\frac{\rho r}{R^2}\right)^{\mu_k},$$

$$\Omega_2^+ = \sum_{k=1}^{\infty}\left[-\frac{\lambda}{2\mu_k}e^{ik(\varphi-\gamma)} \frac{P_{-k}^2}{2\bar{\lambda}\mu_k}e^{-i(\varphi-\gamma)}\right]\left(\frac{\rho r}{R^2}\right)^{\mu_k}.$$

We introduce the operator S_G^+:

$$S_G^+ f = -\frac{1}{\pi}\iint_{G+}\left[\frac{\Omega_1^+(z,\zeta)}{\zeta}f(\zeta) + \frac{\Omega_2^+(z,\zeta)}{\bar{\zeta}}\overline{f(\zeta)}\right] d\xi\, d\eta.$$

It is easy to verify that S_G^+ is a completely continuous operator from $L_q\overline{(G+)}, q > 2$, in $C(\overline{G}^+)$, and the function $S_G^+ f$ belongs to the class $C(\overline{G}^+) \cap C_{\bar{z}}(G^+ - O)$ and satisfies the model equation (4.9).

Now let $|z| > R$ and $|\zeta| < R$. We introduce two function pairs

$$\overset{1}{\Omega}{}_1^- = -\frac{1}{2}\left(\frac{\rho}{R}\right)^{|\lambda|} - \sum_{k=1}^{\infty}\left(\frac{\rho}{R}\right)^{\mu_k}\left(\frac{R}{z}\right)^k e^{ik\gamma};$$

$$\overset{1}{\Omega}{}_2^- = \frac{\lambda}{2|\lambda|}\left(\frac{\rho}{R}\right)^{|\lambda|} + \sum_{k=1}^{\infty}\frac{P_{-k}}{\bar{\lambda}}\left(\frac{\rho}{R}\right)^{\mu_k}\left(\frac{R}{z}\right)^k e^{ik\gamma};$$

$$\overset{2}{\Omega}{}_1^- = \sum_{k=|m|}^{\infty}\left(\frac{\rho}{R}\right)^{\mu_k}\left(\frac{R}{z}\right)^k e^{ik\gamma};$$

$$\overset{2}{\Omega}{}_2^- = \sum_{k=|m|}^{\infty}\frac{P_{-k}}{\bar{\lambda}}\left(\frac{\rho}{R}\right)^{\mu_k}\left(\frac{R}{z}\right)^k e^{ik\gamma}.$$

Let us examine the function

$$\overset{p}{S}{}_G^- f = -\frac{2z^{-m}}{\pi}\iint_{G+}\left[\frac{\overset{p}{\Omega}{}_1^-(z,\zeta)}{\zeta}f(\zeta) + \frac{\overset{p}{\Omega}{}_2^-(z,\zeta)}{\bar{\zeta}}\overline{f(\zeta)}\right] d\xi\, d\eta, \quad (p = 1, 2).$$

It is obvious that $\overset{1}{S}{}_G^- f$ with $m \geq O$ and $\overset{2}{S}{}_G^- f$ with $m < O$ are analytic functions in the domain $|z| \geq R, \overset{1}{S}{}_G^- f$ having a zero of order m at infinity and $\overset{2}{S}{}_G^- f$ being bounded.

7.2 We proceed directly to the solution of the problem. We turn to integral equation (4.2) from ch.1:

$$w^+(z) = \Phi^+(z) + S_G F + P_G \overline{w^+}, \quad z \in G. \tag{7.3}$$

On the basis of (7.3) a new integral equation is constructed. To this end we represent

$$\Phi^+(z) = \Phi_O^+(z) + S_G^+ F + P_G^+ \overline{w^+}, \quad z \in G,$$

where

$$P_G^+ \overline{w^+} = S_G^+ \left(\frac{b(\zeta) - b(O)}{2\bar{\zeta}} \overline{w^+(z)} \right)$$

and $\Phi_O^+(z)$ is a new unknown solution of (4.9). Introducing $\Phi^+(z)$ into (7.3) and denoting $\breve{S}_G = S_G + S_G^+$ and $\breve{P}_G = P_G + P_G^+ g$, we get

$$w^+(z) = \Phi_O^+(z) + \breve{S}_G F + \breve{P}_G \overline{w^+}, \quad z \in G^+. \tag{7.4}$$

Simultaneously we write

$$\Phi^-(z) = \Phi_O^-(z) + \overset{p}{S}_G F + \overset{p}{P}{}_G^- \overline{w^+}, \quad z \in G^-, \quad p = 1, 2, \tag{7.5$_p$}$$

where $\Phi_O^-(z)$ is a new analytic function and

$$\overset{p}{P}{}_G^- \overline{w^+} = \overset{p}{S}{}_G^- \left(\frac{b(\zeta) - b(O)}{2\bar{\zeta}} \overline{w^+(z)} \right).$$

The following analysis depends on the sign of m.

7.3 Let $m \geq O$. Introducing $w^+(z)$ and $\Phi^-(z)$ from (7.4) and (7.5$_1$) into (7.2), we get

$$\Phi_O^+(z) = z^m \Phi_O^-(z) + g(z), \quad z \in \Gamma. \tag{7.6}$$

Thus, the pair of functions $w^+(z)$ and $\Phi^-(z)$ will automatically satisfy boundary condition (7.2), if $\Phi_O^+(z)$ and $\Phi_O^-(z)$ are solutions of problem (7.6). The solvability picture of the latter has been described in Theorem 10.1 from ch. 2. We take from there the expressions for $\Phi_o^{\pm}(z)$ and introduce them into (7.4) and (7.5$_1$). Summarizing the solution of the conjugation problem leads to the determination $\Phi^-(z)$ from (7.5$_1$) and the discovery of $w^+(z)$ from integral equation (7.4) in which the right side contains $2m + 1$ arbitrary real constants.

Theorem 7.1 *Let $m \geq O$ and $\|P_G\|_C < 1$. Then the homogeneous problem ($F = g = O$) has $2m+1$ linearly independent solutions, and the inhomogeneous problem is solvable for any* $F(z)$ and $h(z)$.

7.4 Let $m < O$. For solvability of the problem it is necessary for $w^+(z)$ to satisfy the conditions:

$$T_k^*\left(F(\zeta) + \frac{b(\zeta) - b(O)}{2\bar{\zeta}}\overline{w^+(z)}\right) = g_{-k} - \frac{P_{-k}}{\bar{\lambda}}\bar{g}_k, \quad k = 0, 1, \dots, |m| - 1, \tag{7.7}$$

where

$$T_k^*(f(\zeta)) = \frac{1}{\pi}\iint\limits_{G+}\left[\frac{1}{\zeta}f(\zeta) - \frac{1}{\bar{\zeta}}\frac{P_{-k}}{\bar{\lambda}}\overline{f(\zeta)}\right]\left(\frac{\rho}{R}\right)^{\mu_k} \mathrm{e}^{ik\gamma}\mathrm{d}\xi\,\mathrm{d}\eta.$$

In fact, we substitute the right side of (7.4) for w^+ into (7.2). Then $\Phi_o^+(z)$ and $\Phi_o^-(z)$ form a solution of problem (10.1), ch. 2 and instead of $g(z)$ we should write

$$g(z) - \check{P}_G\overline{w^+} - \check{S}_G F, \quad z \in \Gamma.$$

Now we can verify that (7.7) is a consequence of the solvability conditions (10.9) from ch. 2.

Introducing the expressions $w^+(z)$ and $\Phi^-(z)$ from (7.4) into (7.5_2) and using (7.7), we obtain $\Phi_o^{\pm}(z)$ satisfying the boundary problem described in Consequence 10.1, ch. 2, and having the only solution. Thus, problem (7.1), (7.2) for $m < 0$ is equivalent to finding $w^+(z)$ from integral equation (7.4), the additional conditions (7.7) for $w^+(z)$ and determining $\Phi^-(z)$ by formula (7.5_2).

Theorem 7.1 continuation *Let $m < O$ and $\|\check{P}_G\|_C < 1$. Then the homogeneous problem* (7.1), (7.2) ($F = g = 0$) *has only the zero solution, and for the solvability of the inhomogeneous problem it is necessary and sufficient that* $F(z)$ and $g(z)$ satisfy $2|m| - 1$ real conditions.

The above conditions will issue from (7.7) if instead of $w^+(z)$ we substitute the solutions of equation (7.4). We should also note that the condition $\|\check{P}_G\|_C < 1$ in the simplest cases is realized either for 'small' $b(z)$, or for $b(z)$ differing slightly from $b(0)$, or for mall measure of the domain G^+.

8 The method of constructing additional operators

In §§4–7 the additional operators for investigating the problems of Riemann–Hilbert and linear conjugation have been applied. The awkwardness of their expressions implies that they could be revealed by quite a concrete method and not in a heuristic way. As the process of constructing any operators is interesting in itself and can be utilized in investigating varous problems, in this section the idea of such a construction will be stated in a concise way. It should be noted that all the additional operators of chapters 4 and 5 have been obtained using this idea.

8.1 Additional operators in the Riemann–Hilbert problem. To construct these operators we consider the boundary problem in the form

$$\partial_{\bar{z}} w - \frac{\lambda}{2\bar{z}}\bar{w} = f(z), \quad |z| < R, \tag{8.1}$$

$$\mathrm{Re}(z^{-m} w) = 0 \quad \text{on } |z| = R, \tag{8.2}$$

i.e. one searches for a solution of the inhomogeneous model equation with the homogeneous boundary condition.

According to the results of ch.1 the general solution of equation (8.1) from the class $C(\bar{G}) \cap C_{\bar{z}}(G - O)$ is given by the formula

$$w(z) = \Phi(z) + S_G f, \tag{8.3}$$

where $\Phi(z) \in C(\bar{G}) \cap C_{\bar{z}}(G - O)$ is the general solution of the homogeneous equation (8.1) and $S_G f$ is the partial solution of (8.1).

Introducing (8.3) into (8.2), we get

$$\mathrm{Re}(z^{-m}\Phi) = h(z) \quad \text{on } |z| = R, \tag{8.4}$$

in which

$$h(z) = -\mathrm{Re}(z^{-m} S_G f). \tag{8.5}$$

Thus, in respect of $\Phi(z)$ we come to the Riemann–Hilbert boundary problem (8.4) with the special value of the right side $h(z)$, defined by equality (8.5).

Now we turn to §9, ch. 2, in which this problem was solved. To use this result it is necessary to calculate the Fourier coefficients of the function $h(z)$. First of all we have

$$S_G f(R\mathrm{e}^{\mathrm{i}\varphi}) = \sum_{-\infty}^{\infty} s_k \mathrm{e}^{\mathrm{i}k\varphi}, \tag{8.6}$$

where

$$s_k = \frac{1}{\pi} \iint\limits_{|\zeta| \le R} \left\{ \frac{P_{-k}}{2\mu_k} \frac{f(\zeta)}{\zeta} - \frac{\lambda}{2\mu_k} \frac{\overline{f}(\zeta)}{\bar{\zeta}} \right\} \left(\frac{\rho}{R} \right)^{\mu_k} \mathrm{e}^{-\mathrm{i}k\gamma} d\xi \, d\eta. \tag{8.7}$$

The essence of this notation is explained in §1, ch. 1. We note that equalities (8.6), (8.7) issue from (1.11), (1.12), ch. 1; more precisely, from the second lines in the expressions of the kernels $\Omega_1(z,\zeta)$ and $\Omega_2(z,\zeta)$, since $|z| = R > |\zeta| = \rho$.

From (8.5) by virtue of (8.6) we further get

$$h(R\mathrm{e}^{\mathrm{i}\varphi}) = -R^{-m}\mathrm{Re}\left(\sum_{-\infty}^{\infty} s_{m+k}\mathrm{e}^{\mathrm{i}k\varphi} \right).$$

Hence we determine the Fourier coefficients of the function $h(R\mathrm{e}^{\mathrm{i}\varphi})$:

$$h_k = \frac{1}{2\pi} \int_0^{2\pi} h(R\mathrm{e}^{\mathrm{i}\varphi})\mathrm{e}^{-\mathrm{i}k\varphi}\, d\varphi = -R^{-m} \frac{s_{m+k} + \bar{s}_{m-k}}{2} \quad (k = 0, \pm 1, \ldots). \tag{8.8}$$

Now it is necessary to distinguish the cases $m = 0, m > 0$ and $m < 0$. For any of them in subsections 9.1, 9.2 and 9.3 of ch. 2 the solutions $\Phi(z)$ of problem (8.4) is written out through the Fourier coefficients of the function $h(R\mathrm{e}^{\mathrm{i}\varphi})$. Substituting the right side of (8.8) for h_k into the above solutions and transforming the expressions obtained with regard to (8.7) to a form analogous to S_G (see (1.1) from §1, ch. 1) we get additional operators and kernels that have been used for the solution of the Riemann–Hilbert problem in §§4–6.

8.2 Additional operators in the conjugation problem. For constructing these operators we should consider the boundary value problem:

$$\partial_{\bar{z}} w^+ - \frac{\lambda}{2\bar{z}} \overline{w^+} = f(z), \quad |z| \le R, \tag{8.9}$$

$$w^+(z) = z^m \Phi^-(z) \quad \text{on } |z| = R, \tag{8.10}$$

where $w^+(z) \in C(\overline{G^+}) \cap C_{\bar{z}}(G^+ - 0), G^+ = \{z : |z| < R\}$, and $\Phi^-(z)$ is an analytic function in $G^- = \{z : |z| > R\}$, bounded at $z = \infty$ and continuously extendable on the circle $\Gamma = \{z := R\mathrm{e}^{\mathrm{i}\varphi}\}$. In addition, $f(z) \in L_p(\overline{G^+}), p > 2$.

Just as before, we return to the general solution of equation (8.9):

$$w^+(z) = \Phi^+(z) + S_G f,$$

in which $\Phi^+(z)$ is the general solution of homogeneous problem (8.9). Introducing $w^+(z)$ into the left side of (8.10), we have

$$\Phi^+(z) = z^m \Phi^-(z) + g(z), \quad |z| = R, \tag{8.11}$$

where

$$g(z) = -S_G f(z). \tag{8.12}$$

The solution of problem (8.11), which is contained in §10, ch. 2, is expressed through the Fourier coefficients of the function $g(z) = g(Re^{i\varphi})$. On account of (8.12) and (8.6) we have

$$g_k = \frac{1}{2\pi} \int_O^{2\pi} g(Re^{i\varphi}) e^{-ik\varphi} d\varphi = s_k \quad (k = 0, \pm 1, \ldots),$$

where s_k is given by formula (8.7).

Substituting the expression g_k in the solutions $\Phi^+(z)$ and $\Phi^-(z)$ and changing them to a form similar to the one which has the operator S_G (see (1.1) from §1, ch. 1), we obtain the additional operators $\overset{p}{S}{}^+_G, \overset{p}{S}{}^-_G, (p = 1, 2)$ and kernels $\Omega_1^+, \Omega_2^+, \overset{p}{\Omega}{}_1^-, \overset{p}{\Omega}{}_2^- (p = 1, 2)$, which have been used in §7 of the present chapter for studying the conjugation problem.

Remark In the problems of Riemann–Hilbert and linear conjugation with $m \geq O$ their solutions contain summands with arbitrary constants. It should be noted that these summands do not construct the above additional operators.

9 Unsolved problems

In §§4–7 the study of the boundary problems has been done on the assumption that the corresponding operators satisfy the contractness conditions, which are certainly realized either for 'small' $b(z)$ or for $b(z)$ differing slightly from $b(0)$, or for 'small measure' of the domain G.

In this connection the following question arises: *is it possible to clarify the complete situation for solving integral equations (4.13), (5.3), (6.5) using less restrictive suppositions?*

Chapter 4
Modified generalized Cauchy–Riemann systems with a singular point

An equation of the type

$$\partial_{\bar z} w - \frac{b(z)}{2\bar z} e^{in\varphi} \bar w = F(z), \quad z \in G,$$

(n is an integer different from zero and $b(z), F(z)$ are functions with the properties formulated in the Introduction to ch.3) will be called **modified** with respect to the general case of ch. 3. The novelty of this equation arises from the presence in the coefficient of $\bar w$ of the factor $e^{in\varphi}$, which has a bounded break at the point $z = 0$.

The method of constructing the kernels of the general integral operator, establishing the connection between the sets of solutions for the general and model equations etc. which have been developed in ch. 3, is applied to the present case although the presence of the factor $e^{in\varphi}(n \neq 0)$ specifies the character of the investigation. For this purpose the material of this chapter will be presented in a concise form, the main attention being paid to formulating the final results. More detailed information can be found in [3, 4, 61].

1 The general integral equation

1.1 The development of a theory for the modified generalized Cauchy–Riemann system starts by constructing the general solution of the equation

$$\partial_{\bar z} w - \frac{\lambda}{2\bar z} e^{in\varphi} \bar w = f(z), \quad z \in G, \tag{1.1}$$

where $w(z)$ is the sought function, $f(z)$ is a given function, $\lambda = \lambda_1 + i\lambda_2$ is a complex number and n is an integer; G is a 1–connected domain containing the point $z = 0$ in its interior.

The construction of the general solution and the general integral operator is realized by the scheme described in §1, ch. 1. We deal here with two equations in contrast to the system of differential equations obtained there for seeking the Fourier coefficients $w_k(r), k = 0, \pm 1, \ldots$, of the unknown function $w(z)$:

$$\frac{dw_k}{dr} - \frac{k}{r}w_k - \frac{\lambda}{r}\bar{w}_{n-k} = f_k(r),$$

$$\frac{dw_{n-k}}{dr} - \frac{n-k}{r}w_{n-k} - \frac{\lambda}{r}\bar{w}_k = f_{n-k}(r).$$

Hence it follows that for $f(z) = 0$ the combination

$$w_k(r)\mathrm{e}^{\mathrm{i}k\varphi} + w_{n-k}(r)\mathrm{e}^{\mathrm{i}(n-k)\varphi}$$

is a partial solution of the homogeneous equation (1.1). This implies that the Fourier series of $w(z)$ used for constructing the general solution of equation (1.1) in the present case can be written in the following form

$$w(z) = \sum_{k=\left[\frac{n+1}{2}\right]}^{\infty} \left\{w_k(r)\mathrm{e}^{\mathrm{i}k\varphi} + w_{n-k}(r)\mathrm{e}^{\mathrm{i}(n-k)\varphi}\right\},$$

where square brackets denote the integer part of a number.

Omitting all calculations similar to those done in ch. 1, (there are no significant new difficulties) we eventually get the general solution of equation (1.1) written down as before in the form

$$w(z) = \Phi(z) + S_G f, \quad z \in G. \tag{1.2}$$

The difference is that $\Phi(z)$ is the general solution of the model equation

$$\partial_{\bar{z}}\Phi - \frac{\lambda}{2\bar{z}}\mathrm{e}^{\mathrm{i}n\varphi}\bar{\Phi} = 0, \quad z \in G, \tag{1.3}$$

and the kernels Ω_1 and Ω_2 in operator S_G are expressed by different formulas from those in (1.12), ch.1.

To show their explicit form we introduce the notation:

$$k = 0, \pm 1, \ldots,$$

$$\nu_k^2 = (n-2)^2 + 4|\lambda|^2, \qquad 2\mu_{1,k} = n + \nu_k, \qquad 2\nu_{2,k} = n - \nu_k,$$

$$P_{1,k} = \mu_{1,k} - k, \quad P_{2,k} = \mu_{2,k} - k;$$

$$\mathcal{I}_k(\varphi - \gamma) = \frac{\lambda}{\nu_k}\mathrm{e}^{\mathrm{i}k(\varphi-\gamma)} - \frac{P_{1,k}P_{2,k}}{\bar{\lambda}, \nu_k}\mathrm{e}^{\mathrm{i}(n-k)(\varphi-\gamma)},$$

$$J_k(z,\zeta) = \left[\frac{P_{2,k}}{\nu_k}\left(\frac{r}{\rho}\right)^{\mu_{1,k}} - \frac{P_{1,k}}{\nu_k}\left(\frac{r}{\rho}\right)^{\mu_{2,k}}\right] e^{ik(\varphi-\gamma)}$$
$$+\left[\frac{P_{2,k}}{\nu_k}\left(\frac{r}{\rho}\right)^{\mu_{2,k}} - \frac{P_{1,k}}{\nu_k}\left(\frac{r}{\rho}\right)^{\nu_{1,k}}\right] e^{i(n-k)(\varphi-\gamma)},$$

where $z = re^{i\varphi}$ and $\zeta = \rho e^{i\gamma}$.

Depending on the values of n and $|\lambda|$, it is necessary to distinguish the following four cases. We shall say that the case c_{11} is satisfied and shall denote $(n,\lambda) \in c_{11}$, if $n \leq -1$ and $2|\lambda| = n$. In the same way, $(n,\lambda) \in c_{21}$, if $n \leq -1$ and $2|\lambda| > n$; $(n,\lambda) \in c_{12}$, if $n \geq 1$ and $2|\lambda| \leq n$; $(n,\lambda) \in c_{22}$, if $n \geq 1$ and $2|\lambda| > n$.

Let

$$k_O = \left[\frac{|n| - \sqrt{|n|^2 - 4|\lambda|^2}}{2}\right] \quad \text{for } (n,\lambda) \in c_{11}$$

and

$$k_O^* = \left[\frac{n + \sqrt{|n|^2 - 4|\lambda|^2}}{2}\right] \quad \text{for } (n,\lambda) \in c_{12}.$$

In addition, let $p = -k_O$ in case c_{11}; $p = -\left[\frac{|n|}{2}\right]$ in case c_{21}; $p = k_O^* + 1$ in case c_{12} and $p = [\frac{n+1}{2}]$ in case c_{22}. In this paragraph square brackets always denote the integer part of a number.

Now we have an opportunity to define the function Ω_1 and Ω_2:

$$\Omega_1 = \begin{cases} \omega_O + \sum\limits_{k=p}^{\infty} \dfrac{P_{2,k}e^{i(n-k)(\varphi-\gamma)} - P_{1,k}e^{ik(\varphi-\gamma)}}{\nu_k}\left(\dfrac{r}{\rho}\right)^{\mu_{2,k}}, |\zeta| < |z|, \\ \omega_{OO} + \sum\limits_{k=p}^{\infty} \dfrac{P_{1,k}e^{i(n-k)(\varphi-\gamma)} - P_{2,k}e^{ik(\varphi-\gamma)}}{\nu_k}\left(\dfrac{r}{\rho}\right)^{\mu_{1,k}}, |\zeta| > |z|, \end{cases} \tag{1.4}$$

$$\Omega_2 = \begin{cases} e^{in\gamma}\left(\omega^O + \sum\limits_{k=p}^{\infty} I_k\left(\dfrac{r}{\rho}\right)^{\mu_{2,k}}\right), & |\zeta| < |z|, \\ e^{in\gamma}\left(\omega^{OO} + \sum\limits_{k=p}^{\infty} I_k\left(\dfrac{r}{\rho}\right)^{\mu_{1,k}}\right), & |\zeta| > |z|. \end{cases}$$

Here $\omega_O, \omega_{OO}, \omega^O, \omega^{OO}$ are fully concrete functions of z and ζ. Two of them ω_O and ω^O are different from zero only in c_{11}. The two others, ω_{OO} and ω^{OO} are different from zero only in c_{12}. We write out their expressions when they are not equal to zero.

For $(n, \lambda) \in c_{11}$

$$\omega_O = \sum_{k=-[\frac{|n|}{2}]}^{-k_O-1} J_k(z,\zeta),$$

$$\omega^O = \sum_{k=-[\frac{|n|}{2}]}^{-k_O-1} \left\{\left(\frac{r}{\rho}\right)^{\mu_{2,k}} - \left(\frac{r}{\rho}\right)^{\mu_{1,k}}\right\} I_k(\varphi-\gamma);$$

for $(n, \lambda) \in c_{12}$

$$w_{OO} = -\sum_{[\frac{n+1}{2}]}^{k_O^*} J_k(z,\zeta),$$

$$w^{OO} = -\sum_{[\frac{n+1}{2}]}^{k_O^*} \left\{\left(\frac{r}{\rho}\right)^{\mu_{1,k}} - \left(\frac{r}{\rho}\right)^{\mu_{2,k}}\right\} I_k(\varphi-\gamma).$$

1.2 The functions Ω_1 and Ω_2 possess properties similar to those in equation (1.1) from ch.1; see Lemmas 2.1 and 2.2 of ch.1. Without specifically enumerating these properties, we observe

$$\partial_{\bar z}\Omega_1 - \frac{\lambda}{2\bar z}e^{in\varphi}\bar\Omega_2 = 0, \qquad \partial_{\bar z}\Omega_2 - \frac{\lambda}{2\bar z}e^{in\varphi}\bar\Omega_1 = O.$$

The same situation applies to the operator

$$S_G f = -\frac{1}{\pi}\iint_G \left[\frac{\Omega_1(z,\zeta,n)}{\zeta} f(\zeta) + \frac{\Omega_2(z,\zeta,n)}{\zeta}\overline{f(\zeta)}\right] d\xi\, d\eta,$$

which defines in formula (1.2) a partial solution of equation (1.1) and for which the statements of §3, ch.1 are valid.

These facts create the basis for the detailed investigation of the equation

$$\partial_{\bar z} w - \frac{b(z)}{2\bar z}e^{in\varphi}\bar w = F(z), \quad z \in G, \tag{1.5}$$

in which $w(z) \in C(\bar G) \cap C_{\bar z}(G - O), F(z) \in C(G - O) \cap L_q(\bar G), q > 2$, and the coefficient $b(z)$ is continuous in $\bar G$; in addition $b(0) \neq O$ and

$$|b(z) - b(0)| < M|z|^\alpha, \alpha > 0,$$

for z belonging to an arbitrarily small neighbourhood of the point $z = 0$.

1.3 If we assume $\lambda = b(0)$ and

$$f(z) = F(z) + \frac{b(z) - b(0)}{2\bar{0}} e^{in\varphi} \overline{w(z)},$$

then (1.5) takes the form of equation (1.1). Therefore using formula (1.2) in (1.5) we arrive at an integral equation for $w(z)$:

$$w(z) = \Phi(z) + S_G F + P_G \bar{w}, \quad z \in G, \tag{1.6}$$

$$P_G \bar{w} = S_G \left(\frac{b(z) - b(0)}{2\bar{z}} e^{in\varphi} \overline{w(z)} \right)$$

This integral equation serves as the main tool for studying the modified generalized Cauchy–Riemann system with the singular point (1.5).

The operator P_G and its analogy from §4, ch. 1 have identical properties. In particular, P_G is completely continuous in $C(\bar{G})$ and maps this space to itself. The one–valued solvability of equation (1.6) is formulated on condition that $\|P_G\|_C < 1$. This inequality can be satisfied either for 'small measure' of the domain G, or for $b(z)$ in some sense differing 'only a little' from the value $b(0)$. In this case in the function class $C(\bar{G}) \cap C_{\bar{z}}(G - O)$ a one–to–one correspondence is established between the sets $\{\Phi(z)\}$ and $\{w(z)\}$ of the solutions to equations (1.3) and (1.5).

Comparing these results with those obtained for the singular equation of ch.1, we discover not only their qualitative coincidence but also their possible development in the theory of the solutions to equation (1.5) by the same scheme as in ch. 2 and 3.

2 Elements of the theory of the model equation

The basic tool in analysing equation (1.3) is the generalized Cauchy formula:

$$-\frac{1}{2\pi i} \int_\Gamma \frac{\Omega_1(z,\zeta,n)}{\zeta} \Phi(\zeta)\, d\zeta - \frac{\Omega_2(z,\zeta,n)}{\zeta} \overline{\Phi(\zeta)}\, d\zeta$$

$$= \begin{cases} \Phi(z) & \text{for } z \in G, \\ \frac{\alpha}{2}\Phi(z) & \text{for } z \in \Gamma, \\ 0 & \text{for } z \notin G + \Gamma. \end{cases} \tag{2.1}$$

Hence G is a multiply connected domain of a complex plane, bounded by a finite number of simple smooth closed contours Γ and containing the point $z = 0$ inside itself; $\alpha\pi$ is an interior angle of Γ at the point z; Ω_1 and Ω_2 are the functions defined by (1.4).

The proof of this formula when $\Phi(z) \in C(\bar{G}) \cap C_{\bar{z}}(G-0)$ (also for wider classes of functions) does not contain anything new compared to the one in §2, ch. 2.

2.1 Auxiliary sets of complex numbers A large number of results relating to solutions of equation (1.3) depends on the combination of values of n and $|\lambda|$. To make the following account more convenient we introduce a finite number of auxiliary sets of a complex parameter λ for our examination.

Let $n \neq 0$ be an integer and $\Lambda = (\lambda = \lambda_1 + \mathrm{i}\lambda_2.\ \ 0 < |\lambda| < \infty)$ be the plane of complex numbers without the points $\lambda = 0$ and $\lambda = \infty$. For $n = -2, -3, \ldots$ we assume

$$\Lambda = \bigcup_{s=0}^{[|n|/2]} \Lambda_s^-,$$

where

$$\Lambda_0^- = \{\lambda : 0 < |\lambda| < \sqrt{|n|-1}\},$$

$$\Lambda_s^- = \{\lambda : \sqrt{s(|n|-s)} \leq |\lambda| < \sqrt{(s+1)(|n|-s-1)}\} \quad (s = 1, 2, \ldots [\frac{|n|}{2}] - 1),$$

$$\Lambda^-_{\left[\frac{|n|}{2}\right]} = \{\lambda : \sqrt{s(|n|-s)} \leq |\lambda| < \infty, s = \left[\frac{|n|}{2}\right]\}.$$

For $n = -1, 1, 2, \ldots$ we suppose

$$\Lambda = \bigcup_{s=0}^{[n/2]} \Lambda_s^+$$

where

$$\Lambda_s^+ = \{\lambda : \sqrt{s(n-s)} < |\lambda| \leq \sqrt{(s+1)(n-s-1)}\} \quad (s = 0, 1, \ldots, \left[\frac{|n|}{2}\right] - 1),$$

$$\Lambda^+_{[\frac{|n|}{2}]} = \{\lambda : \sqrt{s(n-s)} < |\lambda| < \infty, s = [\frac{|n|}{2}]\}.$$

It should be noted that the integer part of a number is denoted by square brackets.

2.2 The representation of $\Phi(z)$ through series Let $G = \{z : |z| < R\}$ be a disk and $\Gamma = \{z : |z| = R\}$ be a circle. Assuming $\zeta = Re^{\mathrm{i}\theta}$ and $\Phi(\theta) = \Phi(Re^{\mathrm{i}\theta})$, we rewrite (2.1) with $z \in G$ in the form

$$\Phi(z) = \frac{1}{2\pi} \int_0^{2\pi} \{\Omega_1(z, Re^{\mathrm{i}\theta}, n)\Phi(\theta) + \Omega_2(z, Re^{\mathrm{i}\theta}, n)\overline{\Phi(\theta)}\}\, d\theta.$$

Now let us investigate Ω_1 and Ω_2 in formula (1.4). Substituting their expressions, corresponding to the **inequality** $|z| < |\zeta| = R$, under the integral sign and integrating term–by–term we get

for $n \leq -2$ and $\lambda \in \Lambda_s^-(s = 0, 1, \ldots, [\frac{|n|}{2}])$

$$\Phi(z) = \sum_{k=-s}^{\infty} \left\{ \frac{P_{1,k}\alpha_{n-k} + \lambda \bar{a}_k}{\nu_k} e^{i(n-k)\varphi} + \frac{\lambda \bar{a}_{n-k} - P_{2,k}a_k}{\nu_k} e^{ik\varphi} \right\} r^{\mu_{1,k}};$$

for $n \geq 2$ and $\lambda \in \Lambda_s^+(s = 0, 1, \ldots, [\frac{|n|}{2}] - 1)$

$$\Phi(z) = -\sum_{k=[\frac{n+1}{2}]}^{n-s-1} \left\{ \frac{P_{2,k}b_{n-k} + \lambda \bar{b}_k}{\nu_k} e^{i(n-k)\varphi} + \frac{\lambda \bar{b}_{n-k} - P_{1,k}b_k}{\nu_k} e^{ik\varphi} \right\} r^{\mu_{2,k}};$$

$$+ \sum_{k=[\frac{n+1}{2}]}^{\infty} \left\{ \frac{P_{1,k}\alpha_{n-k} + \overline{\lambda a_k}}{\nu_k} e^{i(n-k)\varphi} + \frac{\lambda \bar{a}_{n-k} - P_{2,k}a_k}{\nu_k} e^{ik\varphi} \right\} r^{\mu_{1,k}};$$

for $n \geq -1$ and $\lambda \in \Lambda^+_{[\frac{n}{2}]}$

$$\Phi(z) = \sum_{k=[\frac{n+1}{2}]}^{\infty} \left\{ \frac{P_{1,k}\alpha_{n-k} + \lambda \bar{a}_k}{\nu_k} e^{i(n-k)\varphi} + \frac{\lambda \bar{a}_{n-k} - P_{2,k}a_k}{\nu_k} e^{ik\varphi} \right\} r^{\mu_{1,k}}.$$

In these formulas we use the notation:

$$a_k = \frac{R^{-\mu_{1,k}}}{2\pi} \int_O^{2\pi} \Phi(\theta) e^{-ik\theta}\, d\theta,$$

$$\tag{2.2}$$

$$b_k = \frac{R^{-\mu_{2,k}}}{2\pi} \int_O^{2\pi} \Phi(\theta) e^{-ik\theta} d\theta, \quad k = 0, \pm 1, \ldots.$$

The continuity of $\Phi(z)$ on Γ provides the uniform convergence in G of the series obtained.

2.3 Regularity of the solution at the singular point It is fairly simple to establish the following statements from the formulas of the previous subsection.

Theorem 2.1 *The solutions $\Phi(z)$ of equation (1.3) from the class $C(\bar{G}) \cap C_{\bar{z}}(G - 0)$ belong to the class of functions analytic with respect to z* and $\bar{z}$ *outside $z = 0$.*

Theorem 2.2 *Any solution $\Phi(z)$ from the class $C(\bar{G}) \cap C_{\bar{z}}(G-0)$ vanishes at the point $z = 0$. The possible orders of the zero are for $n \leq -1$:*

$$\mu_{1,k} = \frac{1}{2}(-|n| + \sqrt{(|n|+2k)^2 + 4|\lambda|^2}), \quad k = -\left[\frac{|n|}{2}\right], -\left[\frac{|n|}{2}\right] + 1, \ldots$$

for $n \geq 1$:

$$\mu_{2,k} = \frac{1}{2}(n - \sqrt{(n-2k)^2 + 4|\lambda|^2}), \quad k = \left[\frac{n+1}{2}\right], \ldots, n-s-1,$$

$$\mu_{1,k} = \frac{1}{2}(n + \sqrt{(n-2k)^2 + 4|\lambda|^2}), \quad k = \left[\frac{n+1}{2}\right], \left[\frac{n+1}{2}\right] - 1, \ldots.$$

Theorem 2.3 *For any natural number m equation* (1.3) *has a solution from the class $C^m(G)$.*

In certain combinations of values of n and λ equation (1.3) admits solutions analytic in respect of z and $\bar{z}$ in G, including the point $z = 0$. But we shall not dwell on proving this assertion.

2.4 The analogy to Liouville's theorem Now we consider equation (1.3) on the whole complex plane E. We denote the class of functions, continuous in any finite part of the plane and continuously differentiable with respect to $\bar{z}$ outside $z = 0$, by $C(E) \cap C_{\bar{z}}(E-0)$.

Theorem 2.4 *If a solution of equation* (1.3) *from the class $C(E) \cap C_{\bar{z}}(E-0)$* is bounded on E, then $\Phi(z) \equiv 0$.

The proof of this theorem is based on the representation of the solution in series form and does not raise any difficulties. It is interesting to note that equation (1.3) can have a non–trivial solution in E if at the point $z = 0$ we replace the condition of the continuity of the solution with a condition on the boundedness of the solution. This situation is, however, valid only for concrete combinations of values of n and λ. It is equation (1.3) that can have one, two or four linearly independent solutions for $|n| \geq 2$ and for λ calculated by the formula

$$|\lambda| = \sqrt{s(|n|-s)} \quad (s = 1, 2, \ldots, \left[\frac{|n|}{2}\right]).$$

The concrete number of solutions depends on the evenness of the n and on λ being an essentially complex or imaginary (real) number as well.

2.5 The analogy to the Laurent series Let equation (1.3) be given in a circular ring $G = \{z : R_O < |z| < R\}$. Any solution $\Phi(z)$ of the class $C(\bar{G}) \cap C_{\bar{z}}(G)$ is represented by means of the Cauchy generalized integral formula:

$$\Phi(z) = \frac{1}{2\pi i} \int_{\Gamma_O \cup \Gamma} \frac{\Omega_1(z,\zeta,n)}{\zeta} \Phi(\zeta)\, d(\zeta) - \frac{\Omega_2(z,\zeta,n)}{\bar{\zeta}} \overline{\Phi(\zeta) d(\zeta)},$$

where Γ_O and Γ are the circles $|z| = R_O$ and $|z| = R$, respectively. Assuming $\zeta = R_O e^{i\theta_O}$ on Γ_O and $\zeta = Re^{i\theta}$ on Γ and inserting the expressions Ω_1 and Ω_2 from (1.4), we get after term–by–term integration

$$\begin{aligned}\Phi(z) = \Psi(z) &+ \sum_{k=p}^{\infty} \left\{ \frac{P_{1,k} a_{n-k} + \lambda \bar{a}_k}{\nu_k} e^{i(n-k)\varphi} + \frac{\lambda \bar{a}_{n-k} - P_{2,k} \bar{a}_k}{\nu_k} e^{ik\varphi} \right\} r^{\mu_{1,k}} \\ &- \sum_{k=q}^{\infty} \left\{ \frac{P_{2,k} c_{n-k} + \lambda \bar{c}_k}{\nu_k} e^{i(n-k)\varphi} + \frac{\lambda \bar{c}_{n-k} - P_{1,k} \bar{c}_k}{\nu_k} e^{ik\varphi} \right\} r^{\mu_{2,k}},\end{aligned} \tag{2.3}$$

where

$$\Psi = \begin{cases} -\displaystyle\sum_{k=n^*}^{n-s-1} \left\{ \frac{P_{2,k} c_{n-k} + \lambda c_k}{\nu_k} e^{i(n-k)\varphi} + \frac{\lambda \bar{c}_{n-k} - P_{1,k} c_k}{\nu_k} e^{ik\varphi} \right\} r^{\mu_{2,k}} \\ \quad \text{for } n \geq 2 \text{ and } \lambda \in \Lambda_s^+,\ s = 0, 1, \ldots, [\frac{n}{2}] - 1; \\ \displaystyle\sum_{k=n^*}^{-s-1} \left\{ \frac{P_{1,k} d_{n-k} + \lambda \bar{d}_k}{\nu_k} e^{i(n-k)\varphi} + \frac{\lambda \bar{d}_{n-k} - P_{2,k} d_k}{\nu_k} e^{ik\varphi} \right\} r^{\mu_{1,k}} \\ \quad \text{for } n \leq -2 \text{ and } \lambda \in \Lambda_s^-,\ s = 0, 1, \ldots, [\frac{|n|}{2}] - 1; \\ 0 \quad \text{for } n \leq -2 \text{ and } \lambda \in \Lambda^-_{[\frac{|n|}{2}]}, \text{ also for } n \geq -1 \text{ and } \lambda \in \Lambda^+_{[\frac{n}{2}]}; \end{cases}$$

a_k, b_k are given by formulas (2.2);

$$c_k = \frac{R_0^{-\mu_{2,k}}}{2\pi} \int_O^{2\pi} \Phi(R_O e^{i\theta_O}) e^{-ik\theta} d\theta_O,$$

$$d_k = \frac{R_O^{-\mu_{1,k}}}{2\pi} \int_O^{2\pi} \Phi(R_O e^{i\theta_O}) e^{-ik\theta_O}\, d\theta_O;$$

$$p = q = \left[\frac{n+1}{2}\right] \text{ for } n \leq -2 \text{ and } \lambda = \Lambda^-_{\left[\frac{|n|}{2}\right]}, \text{ also for } n \geq -1 \text{ and } \lambda \in \Lambda^+_{\left[\frac{n}{2}\right]};$$

$$p = \left[\frac{n+1}{2}\right] \text{ and } q = n - s \text{ for } n \geq 2 \text{ and } \lambda \in \Lambda^+_s (s = 0, 1, \ldots, \left[\frac{n}{2}\right] - 1);$$

$$p = -s \text{ and } q = \left[\frac{n+1}{2}\right] \text{ for } n \leq -2 \text{ and } \lambda \in \Lambda^-_s (s = 0, 1, \ldots, \left[\frac{|n|}{2}\right] - 1).$$

Formula (2.3) defines the expansion of solutions to equation (1.3) by analogy to the Laurent series.

2.6 Supplementary information on properties of the solution of the model equation (1.3) issues from the generalized integral of Cauchy type:

$$\Phi(z) = \frac{1}{2\pi i} \int_\Gamma \frac{\Omega_1(z, \zeta, n)}{\zeta} \nu(\zeta)\, d(\zeta) - \frac{\Omega_2(z, \zeta, n)}{\bar{\zeta}} \overline{\nu(\zeta) d(\zeta)}.$$

Here Γ is a simple smooth closed contour dividing the complex plane E on the domains $G^+ (z = 0 \in G^+)$ and $G^- (z = \infty \in G^-)$; Ω_1 and Ω_2 are the functions given by formulas (1.4), and $\nu(\zeta)$ is a function satisfying the Holder condition on Γ.

This integral determines a pair of functions: $\Phi^+(z), z \in G^+$, and $\Phi^-(z), z \in G^-$; either of them is a solution of equation (1.3). Moreover $\Phi^+(z) \in C(G^+ + \Gamma) \cap C_{\bar{z}}(G^+ - O)$ and $\Phi^-(z) \in C(G^- + \Gamma) \cap C_{\bar{z}}(G^-)$. Limit values of these functions are connected on the integration contour by Sohockii–Plemelj formulas.

As in §§7,8 ch. 2, we can analyse the problem of the continuous extension of a function $\nu(\zeta), \zeta \in \Gamma$, into a solution in G^+ (or G^-) of equation (1.3) and can clarify the cases of a one–to–one correspondence between the set of solutions $\{\Phi(z)\}$ of equation (1.3) in G^+ (or G^-) and the set of functions $\{\nu(\zeta)\}$ given on Γ.

3 Boundary value problems for solutions of the model equation

In this section we present the results of the Riemann–Hilbert boundary problem and the problem of linear conjugation. We omit the corresponding proofs as they do not contain any novelty in comparison with §§9, 10 ch. 2.

3.1 The Riemann–Hilbert problem *It is necessary to find solutions of equation (1.3) in a disk $G = \{z : |z| \leq R\}$ which satisfy on the circle $\Gamma = \{z : |z| \leq R\}$ the condition*

$$\operatorname{Re}(z^{-m}\Phi) = h(z).$$

One assumes that m is an integer; $\Phi(z)$ is continuous in $\bar{G}-0$, continuously differentiable with respect to $\bar{z}$ in $G-0$ and bounded at $z=0$; the given function $h(z)=h(Re^{i\varphi})$ is separable by the uniformly and absolutely convergent Fourier series

$$h(Re^{i\varphi}) = h_O + \frac{1}{2}\sum_{k=1}^{\infty} h_k e^{ik\varphi} + \bar{h}_k e^{-ik\varphi},$$

where

$$h_O = \frac{1}{2\pi}\int_O^{2\pi} h(\varphi)\,d\varphi, \qquad h_k = \frac{1}{\pi}\int_O^{2\pi} h(\varphi)e^{-ik\varphi}\,d\varphi, \quad (k=1,2,\ldots).$$

Theorem 3.1 **1.** *Let* $n \le -2$, $\lambda \in \Lambda^-_{\left[\frac{|n|}{2}\right]}$ *or* $n \ge -1$, $\lambda \in \Lambda^+_{\left[\frac{n}{2}\right]}$. *Then for* $n < 2m$ *the homogeneous problem* ($h=0$) *has* $2m-n$ *linearly independent solutions and the inhomogeneous problem is definitely solvable.*

For $n = 2m$ with $\lambda < O$ *the homogeneous problem has only a non–trivial solution and the inhomogeneous problem is solvable if a single condition is fulfilled. For other* λ *the homogeneous problem has only the zero solution and the inhomogeneous problem is always solvable.*

For $n > 2m$ *the homogeneous problem has only the trivial solution and the inhomogeneous problem is solvable if* $n-2m$ *conditions are fulfilled.*

2. *Let* $n \le -2$, $\lambda \in \Lambda^-_s$ $(s = 0,1,\ldots,\left[\frac{|n|}{2}\right]-1)$. *Then for* $m \ge -s$ *the homogeneous problem has* $2(m+s)+1$ *linearly independent solutions and the inhomogeneous problem is always solvable.*

For $m < -s$ *the homogeneous problem has only the zero solution and the inhomogeneous problem is solvable if* $2(|m|-s)-1$ *conditions are fulfilled.*

3. *Let* $n \ge 2, \lambda \in \Lambda^+_s$ $(s=0,1,\ldots,\left[\frac{n}{2}\right]-1)$. *Then for* $m > s$ *the homogeneous problem has* $2(m+s)-1$ *linearly independent solutions and the inhomogeneous problem is always solvable.*

For $m \le s$ *the homogeneous problem has only the zero solution and the inhomogeneous problem is solvable if* $2(s-m)+1$ *conditions are fulfilled.*

The linearly independent solutions of the homogeneous problem and the solvability conditions of the inhomogeneous problem, mentioned in this theorem, are expressed by bulky formulas [4] and therefore are not written down here.

3.2 The problem of linear conjugation Let $\Phi^+(z)$ be a solution of equation (1.3) in the domain $G^+ = \{z : |z| < R\}$, continuous in $\bar{G}^+ - 0$, continuously differentiable in respect of $\bar{z}$ in $G^+ - 0$ and bounded at $z = 0$. Let $\Phi^-(z)$ be an analytic function in $G^- = \{z : |z| > R\}$, bounded at $z = \infty$ and continuously extendable on $\Gamma = \{z : |z| = R\}$.

It is necessary to seek functions $\Phi^+(z), z \in G^+$, and $\Phi^-(z), z \in G^-$, connected on Γ by the condition

$$\Phi^+(z) = z^m \Phi^-(z) + g(z), \quad z \in \Gamma,$$

where m is an integer and $g(z)$ is a given function expanded in an absolute and uniformly convergent Fourier series

$$g(Re^{i\varphi}) = g_0 + \sum_{k=1}^{\infty} g_k e^{ik\varphi} + g_{-k} e^{-ik\varphi},$$

$$g_k = \frac{1}{2\pi} \int_0^{2\pi} g(Re^{i\varphi}) e^{-ik\varphi} \, d\varphi, \quad k = 0, \pm 1, \dots .$$

Theorem 3.2 **1.** *Let $n \leq -2$, $\lambda \in \Lambda^-_{\left[\frac{|n|}{2}\right]}$ or $n \geq -1$, $\lambda \in \Lambda^+_{\left[\frac{n}{2}\right]}$. Then for $n < 2m+1$ the pair of functions $\Phi^+_O(z)$ and $\Phi^-_O(z), z \in G^\pm$, constituting a solution of the homogeneous problem $(g(z) \equiv O)$, contains $2m - n + 1$ arbitrary real constants and the inhomogeneous problem $(g(z) \neq O)$ is undoubtedly solvable.*

For $n \geq 2m+1$ *the homogeneous problem has only the zero solution and the inhomogeneous problem is solvable if $n - 2m - 1$ real conditions are satisfied.*

2. *Let $n \leq -2$, $\lambda \in \Lambda^-_s (s = 0, 1, \dots, \left[\frac{|n|}{2}\right] - 1)$. Then for $m > -s-1$ the solution of the homogeneous problem contains $2(m+s+1)$ arbitrary real constants and the inhomogeneous problem is certainly solvable.*

For $m \leq -s-1$ the homogeneous problem has only the zero solution and the inhomogeneous problem is solvable if $-2(m+s+1)$ real conditions are satisfied.

3. *Let $n \geq 2$, $\lambda \in \Lambda^+_s (s = 0, 1, \dots, [\frac{n}{2}] - 1)$. Then for $m > s$ the solution of the homogeneous problem has $2(m-s)$ arbitrary real constants and the inhomogeneous problem is always solvable.*

For $m \leq s$ the homogeneous problem has only the zero solution and the inhomogeneous problem is solvable if $2(s-m)$ real conditions are satisfied.

As in the previous subsection here we do not extract the bulky expressions for the linearly independent solutions of the homogeneous problem and the solvability conditions of the inhomogeneous problem [4].

4. The general equation

Now we turn our attention to the equation

$$\partial_{\bar{z}} w - \frac{b(z)}{2\bar{z}} e^{in\varphi} \overline{w} = F(z), \quad z \in G. \tag{4.1}$$

Having at hand the theory of model equation (1.3) we can proceed to study equation (4.1) in the same way as has been done in the case $n = 0$ of ch. 3.

In this way the integral equation (1.6) becomes an effective means for modelling the properties of solutions to equation (4.1) through the similar methods of finding solutions to equation (1.3). In particular, we establish that solutions of equations (4.1) for $F = 0$ and (1.3) possess identical structure of zeros at the point $z = 0$. Moreover, the solutions with identical asymptotic behaviour in a neighbourhood of the singular point prove to be in one–to–one correspondence. It will be justified if the concrete condition of smallness (either with respect to $b(z)$, or $b(z) - b(0)$, or the measure of the initial domain) is satisfied. In similar conditions the solvability pictures of boundary value problems for the model equation and the general equation coincide. If we restrict ourselves to complete continuity conditions of the operators then we must satisfy ourselves with calculating indices for the boundary problem. They are, as expected, equal to the indices of the corresponding boundary problem for a model equation which are computed from Theorems 3.1 and 3.2 as the difference of the number of linearly independent solutions of the homogeneous problem and the number of solvability conditions of the inhomogeneous problem.

We do not give any proofs to justify these statements as they do not contain anything new in comparison to the methods used in §§4–7 ch. 3 in studying similar questions for $n = 0$.

It remains to say that these problems are unsolved with respect to equation (4.1), which were formulated in §9, ch. 3 for the simplest equation.

Chapter 5
Generalized Cauchy–Riemann system with the order of the singularity at a point strictly greater than 1

The present chapter is devoted to investigating the complex equation

$$\partial_{\bar z} w - \frac{b(z)}{2r^{1+2\nu}} e^{i(n+1)\varphi} \bar{w} = F(z), \quad z \in G,$$

where G is a simply connected domain of a complex plane $z = x + iy = re^{i\varphi}$, containing the point $z = O$ in its interior and bounded by a simple closed Ljapunov contour L, n is an integer and ν is a positive number; $w(z)$ is the sought function and $b(z)$ and $F(z)$ are given functions. One assumes that $b(z)$ is continuous in $G = G + L, b(O) \neq O$ and

$$|b(z) - b(O)| < M|z|^{6\nu+\varepsilon}$$

where M and ε are positive constants, ε being an arbitrary small number.

The development of the theory of the equation with a high–order polar singularity will be done in relation to its simplest representative

$$\partial_{\bar z} \Phi - \frac{b(O)}{2r^{1+2\nu}} e^{i(n+1)\varphi} \bar{\Phi} = O, \quad z \in G,$$

which is the model equation with respect to the initial equation. Since the research methods utilized here mainly duplicate those which have been explained in the previous chapters, their description will be supplemented by appropriate details, especially in those places where the specific features of our object look sufficiently vivid.

It should be noted that the present chapter comprises the results obtained in [41–43].

1 Constructing general solutions of an inhomogeneous model equation

Let us consider the equation

$$\partial_{\bar z} w - \frac{\lambda}{2r^{1+2\nu}} e^{i(n+1)\varphi} \bar{w} = f(z), \quad z \in G, \tag{1.1}$$

in which $w(z)$ is the sought function and $f(z)$ is a given function, $z = re^{i\varphi}, \lambda$ is a complex number and ν is a positive number, and n is an integer.

The aim of this section is to construct the general solution of equation (1.1) in a formal way and to get simultaneously the explicit form of a general integral operator which will often be used in future. The scheme of constructing the operator mainly repeats that in §1. The proper calculation will be presented here fairly completely.

Let $G = \{z : |z| \leq R\}$. We rewrite (1.1) on polar coordinates

$$\frac{\partial w}{\partial r} + \frac{i}{r}\frac{\partial w}{\partial \varphi} - \frac{\lambda e^{in\varphi}}{2r^{1+2\nu}}\bar{w} = 2e^{-i\varphi}f. \tag{1.2}$$

We seek a solution $w(z)$ in Fourier series with respect to φ:

$$w(z) = \sum_{-\infty}^{\infty} w_k(r)e^{ik\varphi},$$

where

$$w_k(r) = \frac{1}{2\pi}\int_O^{2\pi} w(z)e^{-ik\varphi}d\,\varphi.$$

Introducing this series into (1.2) and equating coefficients of the same powers of $e^{ik\varphi}$, we get

$$\frac{\partial w_k}{\partial r} - \frac{k}{r}w_k - \frac{\lambda}{r^{1+2\nu}}\bar{w}_{n-k} = f_k, \quad O \leq r \leq R, \tag{1.3}$$

where

$$f_k(r) = \frac{1}{\pi}\int_0^{2\pi} f(z)e^{-i(k+1)\varphi}\,d\varphi.$$

As k is arbitrary we replace it in (1.3) by $n-k$:

$$\frac{\partial w_{n-k}}{\partial r} - \frac{n-k}{r}w_{n-k} - \frac{\lambda}{r^{1+2\nu}}\bar{w}_k = f_{n-k}. \tag{1.4}$$

Equations (1.3) and (1.4) form a closed system for determining the unknowns w_k and w_{n-k}. This in turn prompts us to use a more convenient notation for the expansion of $w(z)$ in Fourier series, viz.

for even $n = 2n_O$

$$w(z) = w_{n_O}e^{in_O\varphi} + \sum_{n_O+1}^{\infty} w_k e^{ik\varphi} + w_{n-k}e^{i(n-k)\varphi} \tag{1.5}$$

for odd $n = 2n_O - 1$

$$w(z) = \sum_{k=n_O}^{\infty} w_k e^{ik\varphi} + w_{n-k} e^{i(n-k)\varphi}. \tag{1.6}$$

Here and elsewhere n_O is the nearest integer to the left of $(n+1)/2$.

For finding the general solution of (1.3), (1.4), we consider the homogeneous system

$$\frac{\partial w_k^O}{\partial r} - \frac{k}{r} w_k^O - \frac{\lambda}{r^{1+2\nu}} \overline{w_{n-k}^O} = O,$$

$$\frac{\partial w_{n-k}^O}{\partial r} - \frac{n-k}{r} w_{n-k}^O - \frac{\lambda}{r^{1+2\nu}} \overline{w_k^O} = O. \tag{1.7}$$

This system is reduced to the second–order equations:

$$r^2 \frac{\partial^2 w_k^O}{\partial r^2} + (2\nu + 1 - n) r \frac{\partial w_k^O}{\partial r} + \left[-k(2\nu + k - n) - |\lambda|^2 r^{-4\nu}\right] w_k^O = O,$$

$$r^2 \frac{\partial^2 w_{n-k}^O}{\partial r^2} + (2\nu + 1 - n) r \frac{\partial w_{n-k}^O}{\partial r} + \left[-(n-k)(2\nu - k) - |\lambda|^2 r^{-4\nu}\right] w_{n-k}^O = O.$$

Each of them is related to Bessell's equation [46], and their general solutions have the forms:

$$w_k^O(r) = r^{\frac{n}{2}-\nu}\left[A_k I_{|p_k|}\left(\frac{|\lambda|}{2\nu} r^{-2\nu}\right) + B_k K_{p_k}\left(\frac{|\lambda|}{2\nu} r^{-2\nu}\right)\right],$$

$$w_{n-k}^O(r) = r^{\frac{n}{2}-\nu}\left[A_{n-k} I_{|p_{k-1}|}\left(\frac{|\lambda|}{2\nu} r^{-2\nu}\right) + B_{n-k} K_{p_{k-1}}\left(\frac{|\lambda|}{2\nu} r^{-2\nu}\right)\right],$$

where $A_k, B_k, A_{n-k}, B_{n-k}$ are arbitrary complex constants, $p_k = (k+\nu-n/2)2\nu$, $I_p(t)$ and $K_p(t)$ are modified Bessel functions (MacDonald functions), see [24,26,74]. On this basis we get the general solution of homogeneous system (1.7):

$$w_k^O(z) = r^{\frac{n}{2}-\nu}\left[A_k I_{|p_k|}\left(\frac{|\lambda|}{2\nu} r^{-2\nu}\right) + B_k K_{p_k}\left(\frac{|\lambda|}{2\nu} r^{-2\nu}\right)\right],$$

$$w_{n-k}^O(z) = \frac{\lambda}{|\lambda|} r^{\frac{n}{2}-\nu}\left[-\bar{A}_k I_{p_k-1}\left(\frac{|\lambda|}{2\nu} r^{-2\nu}\right) + \bar{B}_k K_{p_k-1}\left(\frac{|\lambda|}{2\nu} r^{-2\nu}\right)\right]. \tag{1.8}$$

Formulas (1.8) are true for both even and odd n. We should especially emphasize the case $n = 2n_O$ and $k = n_O$. In this case both the equations of system (1.7) coincide and the general solution is written out in the following form:

$$w_{n_O}^O(r) = r^{\frac{n}{2}-\nu}\left\{\frac{1}{2}A_{n_O}\left[I_{\frac{1}{2}}\left(\frac{|\lambda|}{2\nu}r^{-2\nu}\right)\right] + I_{-\frac{1}{2}}\left(\frac{|\lambda|}{2\nu}r^{-2\nu}\right) + B_{n_O}K_{\frac{1}{2}}\left(\frac{|\lambda|}{2\nu}r^{-2\nu}\right)\right\}, \tag{1.8^O}$$

the complex constants A_{n_O} and B_{n_O} being connected by the equalities

$$|\lambda|A_{n_O} = -\lambda A_{n_O}, \qquad |\lambda|B_{n_O} = \lambda\bar{B}_{n_O}.$$

Further, using the method of variation of arbitrary constants, we determine the solution of the inhomogeneous system (1.3), (1.4):

$$\begin{aligned} w_k(r) &= r^{\frac{n}{2}-\nu}I_{p_k}\left(\frac{|\lambda|}{2\nu}r^{-2\nu}\right)\int_O^r \frac{\rho^{-\nu-\frac{n}{2}}}{2\nu}\left\{|\lambda|K_{p_k-1}\left(\frac{|\lambda|}{2\nu}\rho^{-2\nu}\right)f_k(\rho)\right.\\ &\left.-\lambda K_{p_k}\left(\frac{|\lambda|}{2\nu}\rho^{-2\nu}\right)\overline{f_{n-k}(\rho)}\right\}\mathrm{d}\rho - r^{\frac{n}{2}-\nu}K_{p_k}\left(\frac{|\lambda|}{2\nu}r^{-2\nu}\right)\\ &\times\int_r^R \frac{\rho^{\nu-\frac{n}{2}}}{2\nu}\left\{|\lambda|I_{p_k-1}\left(\frac{|\lambda|}{2\nu}\rho^{-2\nu}\right)f_k(\rho)\right.\\ &\left.+\lambda I_{p_k}\left(\frac{|\lambda|}{2\nu}\rho^{-2\nu}\right)\overline{f}_{n-k}(\rho)\right\}\mathrm{d}\rho + w_k^O(r), \end{aligned} \tag{1.9}$$

$$\begin{aligned} w_{n-k}(r) &= r^{\frac{n}{2}-\nu}I_{p_k-1}\left(\frac{|\lambda|}{2\nu}r^{-2\nu}\right)\int_O^r \frac{\rho^{-\nu-\frac{n}{2}}}{2\nu}\left\{|\lambda|K_{p_k-1}\left(\frac{|\lambda|}{2\nu}\rho^{-2\nu}\right)\right.\\ &\left.\times f_{n-k}(\rho) - \lambda K_{p_k-1}\left(\frac{|\lambda|}{2\nu}\rho^{-2\nu}\right)\overline{f_k(\rho)}\right\}\mathrm{d}\rho\\ &- R^{\frac{n}{2}-\nu}K_{p_k-1}\left(\frac{|\lambda|}{2\nu}r^{-2\nu}\right)\int_r^R \frac{\rho^{-\nu-\frac{n}{2}}}{2\nu}\left\{|\lambda|I_{p_k}\left(\frac{|\lambda|}{2\nu}\rho^{-2\nu}\right)f_{n-k}(\rho)\right.\\ &\left.+\lambda I_{p_k-1}\left(\frac{|\lambda|}{2\nu}\rho^{-2\nu}\right)\overline{f_k(\rho)}\right\}\mathrm{d}\rho + w_{n-k}^O(r), \end{aligned}$$

where w_k^O and w_{n-k}^O are given by equalities (1.8). For even $n = 2n_O$ and $k = n_O$ on the basis of (1.8) we construct the solution

$$\begin{aligned} w_n(z) &= \frac{1}{2} r^{\frac{n}{2}-\nu}\left[I_{\frac{1}{2}}\left(\frac{|\lambda|}{2\nu}r^{-2\nu}\right) + I_{-\frac{1}{2}}\left(\frac{|\lambda|}{2\nu}r^{-2\nu}\right)\right] \\ &\times \int_O^r \frac{\rho^{-\nu-\frac{n}{2}}}{2\nu} K_{\frac{1}{2}}\left(\frac{|\lambda|}{2\nu}\rho^{-2\nu}\right)\left\{|\lambda| f_{n_O}(\rho) - \lambda\overline{f_{n_O}(\rho)}\right\} \mathrm{d}\rho \\ &- \frac{1}{2} r^{\frac{n}{2}-\nu} K_{\frac{1}{2}}\left(\frac{|\lambda|}{2\nu}r^{-2\nu}\right) \int_r^R \frac{\rho^{-\nu-\frac{n}{2}}}{2\nu}\left[I_{\frac{1}{2}}\left(\frac{|\lambda|}{2\nu}\rho^{-2\nu}\right)\right. \\ &\left. + I_{-\frac{1}{2}}\left(\frac{|\lambda|}{2\nu}r^{-2\nu}\right)\right]\left\{|\lambda| f_{n_O}(\rho) + \lambda\overline{f_{n_O}(\rho)}\right\} \mathrm{d}\rho + w_n^O(z), \end{aligned} \tag{1.9'}$$

where $w_{n_O}^O$ is defined by (1.8^O).

Now in formulas (1.9), (1.9′) the functions $f_{n_O}(\rho), f_k(\rho), f_{n-k}(\rho)$ should be expressed through $f(\zeta), \zeta = \rho e^{ik\psi}$. Then the expressions obtained are introduced either into (1.5) or into (1.6) depending on the evenness of n. Later in a formal way we can interchange the order of integration and summation. Just as in §1, ch. 1, we come to the relation:

$$w(z) = \Phi(z) + S_G f, \quad z \in G, \tag{1.10}$$

Here

$$\begin{aligned} \Phi(z) &= r^{\frac{n}{2}-\nu} \sum_{k=n_O}^{\infty}\left[A_k I_{p_k}\left(\frac{|\lambda|}{2\nu}r^{-2\nu}\right) + B_k K_{p_k}\left(\frac{|\lambda|}{2\nu}r^{-2\nu}\right)\right] e^{ik\varphi} \\ &+ \frac{\lambda}{|\lambda|}\left[-\bar{A}_k I_{p_k-1}\left(\frac{|\lambda|}{2\nu}r^{-2\nu}\right) + \bar{B}_k K_{p_k-1}\left(\frac{|\lambda|}{2\nu}r^{-2\nu}\right)\right] e^{i(n-k)\varphi}, \end{aligned}$$

provided $n = 2n_O$ is an even number. For an odd number n the summands corresponding to indices $k = n_O$ should be divided by 2, in addition

$$|\lambda| A_{n_O} = -\lambda \bar{A}_{n_O} \quad \text{and} \quad |\lambda| B_{n_O} = \lambda \bar{B}_{n_O}.$$

Besides

$$S_G f = -\frac{1}{\pi} \iint_G \left[\frac{\Omega_1(z,\zeta)}{\zeta} f(\zeta) + \frac{\Omega_2(z,\zeta)}{\zeta}\overline{f(\zeta)}\right] \mathrm{d}\xi\, \mathrm{d}\eta \tag{1.11}$$

where $\zeta = \xi + i\eta = \rho e^{i\psi}$, and the functions Ω_1 and Ω_2 form a system of general kernels of equation (1.1). Using the notation

$$H_{\alpha,\beta}(r,\rho) = I_\alpha\left(\frac{|\lambda|}{2\nu}r^{-2\nu}\right) K_\beta\left(\frac{|\lambda|}{2\nu}r^{-2\nu}\right) \tag{1.12}$$

they are defined by the formulas:

$$\Omega_1 = \begin{cases} \dfrac{|\lambda|}{2\nu r^{2\nu}}\left(\dfrac{r}{\rho}\right)^{\nu+\frac{n}{2}} \displaystyle\sum_{k=n_O}^{\infty} \Big\{ H_{p_k-1,p_k}(\rho,r)e^{ik(\varphi-\psi)} \\ +H_{p_k,p_k-1}(\rho,r)e^{i(n-k)(\varphi-\psi)}\Big\} \quad \text{for } |z|<|\zeta|, \\ -\dfrac{|\lambda|}{2\nu r^{2\nu}}\left(\dfrac{r}{\rho}\right)^{\nu+\frac{n}{2}} \displaystyle\sum_{k=n_O}^{\infty} \Big\{ H_{p_k,p_k-1}(r,\rho)e^{ik(\varphi-\psi)} \\ +H_{p_k-1,p_k}(r,\rho)e^{i(n-k)(\varphi-\psi)}\Big\} \quad \text{for } |\zeta|<|z|, \end{cases} \tag{1.13}$$

$$\Omega_2 = \begin{cases} \dfrac{\lambda e^{in\psi}}{2\nu r^{2\nu}}\left(\dfrac{r}{\rho}\right)^{\nu+\frac{n}{2}} \displaystyle\sum_{k=n_O}^{\infty} \Big\{ H_{p_k,p_k}(\rho,r)e^{ik(\varphi-\psi)} \\ +H_{p_k-1,p_k-1}(\rho,r)e^{i(n-k)(\varphi-\psi)}\Big\} \quad \text{for } |z|<|\zeta|, \\ \dfrac{\lambda e^{in\psi}}{2\nu r^{2\nu}}\left(\dfrac{r}{\rho}\right)^{\nu+\frac{n}{2}} \displaystyle\sum_{k=n_O}^{\infty} \Big\{ H_{p_k,p_k}(r,\rho)e^{ik(\varphi-\psi)} \\ +H_{p_k-1,p_k-1}(r,\rho)e^{i(n-k)(\varphi-\psi)}\Big\} \quad \text{for } |\zeta|<|z|, \end{cases}$$

if n is an odd number. For an even n the summands corresponding to indices $k = n_O$, should be divided by 2.

Formula (1.10) is the formally defined general solution of equation (1.1), while $\Phi(z)$ is the general solution of (1.1) for $f(z) \equiv 0$, and $S_G f$ is a partial solution of the inhomogeneous equation (1.1).

2 Properties of the functions Ω_1 and Ω_2

The expressions for $\Omega_k(z,\zeta)$ obtained in the previous section prove to be sufficiently complicated for the following reasons: firstly, they have a different form for $|z| < |\zeta|$ and $|z| > |\zeta|$; secondly, they are constituted by series; and thirdly, the terms of these series are modified Bessel functions which are defined themselves by series. It is quite clear that these reasons generate additional technical obstacles in analysing the properties of the kernels.

2.1 Lemma 2.1 *$\Omega_1(z,\zeta)$ and $\Omega_2(z,\zeta)$ are represented in the form*

$$\Omega_1 = \frac{\zeta}{\zeta - z} + \Omega_1^O(z,\zeta) + \Psi(z,\zeta),$$

$$(2.1)$$

$$\Omega_2 = \Omega_2^O(z,\zeta) - \begin{cases} \lambda r^{in\psi} \ln |1 - \frac{z}{\zeta}|, & |z| \le |\zeta|, \\ \lambda r^{in\psi} \ln |1 - \frac{\zeta}{z}|, & |\zeta| \le |z|, \end{cases}$$

where

$$\Psi(z,\zeta) = \begin{cases} \left.\begin{array}{ll} -\sum_{k=0}^{n_0-1} \left(\frac{r}{\rho}\right)^k e^{ik(\varphi-\psi)}, & |z| \le |\zeta|, \\ 0, & |\zeta| \le |z|, \end{array}\right\} & n_0 > 0, \\ \left.\begin{array}{ll} \sum_{k=0}^{n_0-n-1} \left(\frac{\rho}{r}\right)^k e^{-ik(\varphi-\psi)}, & |\zeta| \le |z|, \\ 0, & |z| \le |\zeta|, \end{array}\right\} & n_0 < 0, \\ 0, & n_0 = 0. \end{cases}$$

For a fixed value z_O ($z_O \ne 0, \infty$) the functions $\Omega_1^O(z_O,\zeta)$ and $\Omega_2^O(z_O,\zeta)$ are *continuous with respect to $\zeta (0 \le |\zeta| < \infty)$, and $\Omega_1(z_O,\zeta)$ and $\Omega_2(z_O,\zeta)$ have a zero of an arbitrarily high order at $\zeta = 0$, and*

$$\Omega_1(z_O,\zeta), \Omega_2(z_O,\zeta) = O(|\zeta|^{-n_O}), \quad \zeta \to \infty. \tag{2.2}$$

In the case when ζ_O ($\zeta_O \ne 0, \infty$) is fixed the functions $\Omega_1^O(z,\zeta_O)$ and $\Omega_2^O(z,\zeta_O)$ are continuous with respect to $z (0 \le z \le \infty)$,and $\Omega_1(z,\zeta_O)$ and $\Omega_2(z,\zeta_O)$ have a zero of an arbitrarily high order at $z = 0$ and

$$\Omega_1(z,\zeta_O), \Omega_2(z,\zeta_O) = O(|z|^{n-n_O}), \quad z \to \infty. \tag{2.2'}$$

Proof Let n be an odd number. We assume $n_0 > O$ for simplicity. Then we fix a value $z = z_O = r_O e^{i\varphi_O}$. Let us establish the continuity of Ω_1^O and Ω_2^O with respect to ζ separately in the domains $|z_O| \le |\zeta| < \infty$ and $O \le |\zeta| \le |z_O|$ and the absence of each discontinuity at $|\zeta| = |z_O|$.

Using the explicit form of Ω_1 and the identity

$$\frac{\zeta}{\zeta - z_O} + \Psi(z_O, \zeta) = \begin{cases} \displaystyle\sum_{k=n_O}^{\infty} \left(\frac{r_O}{\rho}\right)^k e^{ik(\varphi_O - \psi)}, & |z_O| < |\zeta|, \\ \displaystyle -\sum_{k=n+1}^{\infty} \left(\frac{\rho}{r_O}\right)^k e^{i(n-k)(\varphi_O - \psi)}, & |\zeta| < |z_O|, \end{cases}$$

for $|z_O| < |\zeta|$ we get

$$\begin{aligned} \Omega_1^O(z_O, \zeta) = & \sum_{k=n_O}^{\infty} \left\{ \frac{|\lambda|}{2\nu r_O^{2\nu}} \left(\frac{r_O}{\rho}\right)^{\nu + \frac{n}{2}} H_{p_k - 1, p_k}(\rho, r_O) - \left(\frac{r_O}{\rho}\right)^k \right\} e^{ik(\varphi_O - \psi)} \\ & + \sum_{k=n_O}^{\infty} \left\{ \frac{|\lambda|}{2\nu r_O^{2\nu}} \left(\frac{r_O}{\rho}\right)^{\nu + \frac{n}{2}} H_{p_k, p_k - 1}(\rho, r_O) e^{i(n-k)(\varphi_O - \psi)} \right\}. \end{aligned} \tag{2.3}$$

With regard to the properties of modified Bessel functions every term of series (2.3) is a continuous function with respect to ζ for $|z_O| \leq |\zeta|$, therefore to prove the continuity of Ω_1^O it is suffiient to establish the uniform convergence of the series. To this end we estimate the modulus of a term of the series.

First of all we note that

$$\gamma_k(r_O, \rho) = \left(\frac{r_O}{\rho}\right)^k - \frac{|\lambda|}{2\nu r_O^{2\nu}} \left(\frac{r_O}{\rho}\right)^{\nu + \frac{n}{2}} H_{p_k - 1, p_k}(\rho, r_O) \geq O, \quad r_O \leq \rho.$$

Now we consider the expression

$$\begin{aligned} \theta = & \frac{|\lambda|}{2\nu r_O^{2\nu}} \left(\frac{r_O}{\rho}\right)^{\nu + \frac{n}{2}} \left\{ H_{p_k - 1, p_k}(\rho, r_O) + H_{p_k, p_k - 1}(\rho, r_O) \right\} \\ \equiv & \frac{|\lambda|}{2\nu r_O^{2\nu}} \left(\frac{r_O}{\rho}\right)^{\nu + \frac{n}{2}} \left\{ I_{p_k - 1}\left(\frac{|\lambda|}{2\nu} \left(\frac{r_O}{\rho}\right)^{2\nu} r_O^{-2\nu}\right) K_{p_k}\left(\frac{|\lambda|}{2\nu} r_O^{-2\nu}\right) \right. \\ & \left. + I_{p_k}\left(\frac{|\lambda|}{2\nu} \left(\frac{r_O}{\rho}\right)^{2\nu} r_O^{-2\nu}\right) K_{p_k - 1}\left(\frac{|\lambda|}{2\nu} r_O^{-2\nu}\right) \right\}. \end{aligned}$$

Using the properties of modified Bessel functions, see [26, 72],

$$I_p(\beta x) \leq \beta^p I_p(x) \quad (p, x \geq 0, 0 \leq \beta \leq 1) \tag{2.4}$$

$$I_p(x) K_{p+1}(x) + I_{p+1}(x) K_p(x) = \frac{1}{x}, \tag{2.4'}$$

we get $\theta \leq \left(\frac{r_O}{\rho}\right)^k$. Since in the present case $I_p(x), K_p(x) \geq 0$, then it follows that $\gamma_k(r_O, \rho) \geq 0$.

To estimate the maximal value of $\gamma_k(r_O, \rho)$ we compute the derivative

$$\frac{\mathrm{d}\gamma_k}{\mathrm{d}\rho} = -\frac{k}{\rho}\left(\frac{r_O}{\rho}\right)^k + \frac{|\lambda| k}{2\nu\rho r_0^{2\nu}}\left(\frac{r_O}{\rho}\right)^{\nu+\frac{n}{2}} H_{p_k-1,p_k}(\rho, r_O).$$
$$+\frac{|\lambda|\rho^{-2\nu-1}}{2\nu r_O^{2\nu}}\left(\frac{r_O}{\rho}\right)^{\nu+\frac{n}{2}} H_{p_k,p_k}(\rho, r_O).$$

The derivative vanishes at $\rho = \infty$ and takes a positive value at $\rho = r_O$ (this can be checked by direct calculations). Therefore there exists a value $\rho_k (r_O < \rho_k < \infty)$ at which the derivative vanishes and $\gamma_k(r_O, \rho) \leq \gamma_k(r_O, \rho_k)$. Expanding this inequality we arrive at

$$\gamma_k(r_O, \rho) \leq \frac{|\lambda|^2}{2\nu k(\rho_k r_O)^{2\nu}} (\frac{r_O}{\rho})^{\nu+\frac{n}{2}} H_{p_k,p_k}(\rho_k, r_O).$$

Using the first of relations (2.4) and the inequality

$$I_p(x) K_p(x) < \frac{1}{2p} \quad (x, p > 0)$$

we have

$$\gamma_k(r_O, \rho) < \frac{|\lambda|^2 \rho_k^{-4\nu}}{2k(k+\nu-\frac{n}{2})}, \left(\frac{r_O}{\rho_k}\right)^k < \frac{|\lambda|^2 r_O^{-4\nu}}{2k(k+\nu-\frac{n}{2})}, \quad r_O \neq 0.$$

From this estimation the uniform convergence of the first series in (2.3) at a fixed point $z = z_O \neq 0$ follows from the Weierstrass convergence test. For the second series of (2.3) this idea issues from the relation:

$$\frac{|\lambda|}{2\nu r_O^{2\nu}} (\frac{r_O}{\rho})^{\nu+\frac{n}{2}} H_{p_k,p_{k-1}}(\rho_k, r_O) < \frac{|\lambda|^2 r_O^{-4\nu}}{2(k-n)(k+\nu-\frac{n}{2})}.$$

Its correctness is stated by the inequality

$$I_{p+1}(x) K_p(x) < \frac{x}{4p(p+1)}, \quad p, x > 0.$$

For analysing $\Omega_1^O(z_O, \zeta)$ in the domain $|\zeta| < |z_O|$ the following equality should be used

$$\Omega_1^O(z_O,\zeta) = -\frac{|\lambda|}{2\nu r_O^{2\nu}}\left(\frac{r_O}{\rho}\right)^{\nu+\frac{n}{2}}\sum_{k=n_O}^{\infty}\Big\{H_{p_k,p_k-1}(r_O,\rho)e^{ik(\varphi_O-\psi)}$$
$$+ H_{p_k-1,p_k}(r_O,\rho)e^{i(n-k)(\varphi_O-\psi)}\Big\}$$
$$-\frac{|\lambda|}{2\nu r_O^{2\nu}}\left(\frac{r_O}{\rho}\right)^{\nu+\frac{n}{2}}\sum_{k=n+1}^{\infty}H_{p_k,p_k-1}(r_O,\rho)e^{ik(\varphi_O-\psi)}$$
$$-\sum_{k=n+1}^{\infty}\left\{\frac{|\lambda|}{2\nu r_O^{2\nu}}\left(\frac{r_O}{\rho}\right)^{\nu+\frac{n}{2}}H_{p_k-1,p_k}(r_O,\rho)-\left(\frac{\rho}{r_O}\right)^{k-n}\right\}e^{i(n-k)(\varphi_O-\psi)}. \tag{2.5}$$

The continuity of the first sum with respect to ζ for $0 \le |\zeta| \le |z_O|$ is obvious. The proof of the continuity of the other two series with respect to ζ is, just as before, fulfilled by obtaining estimates for every summand and by establishing the uniform convergence of both series.

As the values of $\Omega_1^O(z_O,\zeta)$, taken from (2.3) and (2.5), coincide for $|\zeta| = |z_O|$ then $\Omega_1^O(z_O,\zeta)$ is continuous with $0 \le |\zeta| < \infty$.

Now we prove the continuity of $\Omega_2^O(z,\zeta)$ with respect to ζ for a fixed value $z = z_O \ne 0,\infty$. As in the previous case one establishes the continuity of $\Omega_2^O(z_O,\zeta)$ separately in the domains $|z_O| \le |\zeta| < \infty$ and $0 \le |\zeta| \le |z_O|$ and the absence of discontinuity at $|\zeta| = |z_O|$.

Starting from the identity ($|z_O| < |\zeta|$)

$$\ln|1-\frac{z_O}{\zeta}| \equiv -\sum_{k=1}^{n}\left(\frac{r_O}{\rho}\right)^{k}\frac{e^{ik(\varphi_O-\psi)}}{2k} - \sum_{k=n+1}^{\infty}\left(\frac{r_O}{\rho}\right)^{k-n}\frac{e^{i(n-k)(\varphi_O-\psi)}}{2(k-n)},$$

from equalities (1.13) and (2.1) with $|z| < |\zeta|$ we have

$$\Omega_2^O = \lambda e^{in\psi}\Bigg[\sum_{k=1}^{n_O-1}\left(\frac{r_O}{\rho}\right)^{k}\frac{e^{ik(\varphi_O-\psi)}}{2k}$$
$$+\frac{1}{2\nu r_O^{2\nu}}\left(\frac{r_O}{\rho}\right)^{\nu+\frac{n}{2}}\sum_{k=n_O}^{n}H_{p_k-1,p_k-1}(\rho,r_O)e^{i(n-k)(\varphi_O-\psi)}$$
$$-\sum_{k=n_O}^{\infty}\left\{\frac{1}{2\nu r_O^{2\nu}}\left(\frac{r_O}{\rho}\right)^{\nu+\frac{n}{2}}H_{p_k,p_k}(\rho,r_O)-\frac{1}{2k}\left(\frac{r_O}{\rho}\right)^{k}\right\}e^{ik(\varphi_O-\psi)}$$
$$+\sum_{k=n+1}^{\infty}\left\{\frac{1}{2\nu r_O^{2\nu}}\left(\frac{r_O}{\rho}\right)^{\nu+\frac{n}{2}}H_{p_k-1,p_k-1}(\rho,r_O)\right.$$

$$- \frac{1}{2(k-n)}\left(\frac{r_O}{\rho}\right)^{k-n}\Big\}e^{i(n-k)\varphi_O-\psi}\Big]. \tag{2.6}$$

All summands in (2.6), and moreover the first two sums are continuous functions with $|z_O| \le |\zeta| < \infty$. Let us prove that the other two sums possess the same property. For this purpose we estimate modulus of the summands situated under the symbol of the third and fourth sums. For the summands of the third sum we have

$$\begin{aligned}
&\left|\frac{1}{2\nu r_O^{2\nu}}\left(\frac{r_O}{\rho}\right)^{\nu+\frac{n}{2}} H_{p_k,p_k}(\rho,r_O) - \frac{1}{2k}\left(\frac{r_O}{\rho}\right)^k\right| \\
&\le \left\{\frac{1}{2(k+2\nu)}\left(\frac{r_O}{\rho}\right)^{k+2\nu} - \frac{1}{2\nu r_O^{2\nu}}\left(\frac{r_O}{\rho}\right)^{\nu+\frac{n}{2}} H_{p_k,p_k}(\rho,r_O)\right\} \\
&+\left\{\frac{1}{2k}\left(\frac{r_O}{\rho}\right)^k - \frac{1}{2(k+2\nu)}\left(\frac{r_O}{\rho}\right)^{k+2\nu}\right\}.
\end{aligned}$$

The first bracket of the right–hand inequality proves always to be a non–negative function. It takes its maximal value at a point where its derivative with respect to ρ vanishes. As a result, this bracket is estimated above by the value $|\lambda| r_O^{-4\nu}/2(k+2\nu)(k+\nu-\frac{n}{2})$. As for the second bracket it is estimated by the value $\nu/k(k+2\nu)$.

The uniform convergence of the third series of (2.6) and therefore the continuity of its sum with respect to ζ for a fixed z_O issues from the above estimations.

The proof of the continuity of the function defined by the fourth sum in (2.6) is stated in a similar way.

Quite analaogously the continuity of $\Omega_2^O(z_O,\zeta)$ in the domain $0 \le |\zeta| \le |z_O|$ is proved. In addition we can use the identity:

$$\ln\left|1-\frac{\zeta}{z_O}\right| \equiv -\sum_{k=1}^{n}\left(\frac{\rho}{r_O}\right)^k \frac{e^{i(\varphi_O-\psi)}}{2k} - \sum_{k=n+1}^{\infty}\left(\frac{\rho}{r_O}\right)^{k-n}\frac{e^{i(n-k)(\varphi_O-\psi)}}{2(k-n)},$$

and Ω_2^O should be defined from the equality

$$\begin{aligned}
\Omega_2^O =& \lambda e^{in\psi}\Bigg[\sum_{k=1}^{n_O-1}\left(\frac{\rho}{r_O}\right)^k \frac{e^{ik(\varphi_O-\psi)}}{2k} \\
&+ \frac{1}{2\nu r_O^{2\nu}}\sum_{k=n_O}^{n} H_{p_k-1,p_k-1}(r_O,\rho)e^{i(n-k)(\varphi_O-\psi)} \\
&+ \sum_{k=n_O}^{\infty}\left\{\frac{1}{2\nu r_O^{2\nu}}\left(\frac{r_O}{\rho}\right)^{\nu+\frac{n}{2}} H_{p_k,p_k}(r_O,\rho) - \frac{1}{2k}\left(\frac{\rho}{r_O}\right)^k\right\}e^{ik(\varphi_O-\psi)}
\end{aligned}$$

$$+\sum_{k=n+1}^{\infty}\left\{\frac{1}{2\nu r_O^{2\nu}}\left(\frac{r_O}{\rho}\right)^{\nu+\frac{n}{2}}H_{p_k-1,p_k-1}(r_O,\rho)\right.$$

$$\left.\left.-\frac{1}{2(k-n)}\left(\frac{\rho}{r_O}\right)^{k-n}\right\}e^{i(n-k)(\varphi_O-\psi)}\right]. \tag{2.7}$$

Passing from the domain $O \leq |\zeta| \leq |z_O|$ to the domain $|z_O| \leq |\zeta| < \infty$ the function $\Omega_2^O(z_O,\zeta)$ does not break since values of Ω_2^O, taken from formulas (2.6) and (2.7) at $|\zeta| = |z|$, coincide.

We note that the continuity of $\Omega_1^O(z,\zeta_O)$ and $\Omega_2^O(z,\zeta_O)$ with respect to z at a fixed value $\zeta = \zeta_O \neq O, \infty$ is established according to the previous scheme.

In the same way the continuity of Ω_1^O and Ω_2^O for $n_O \leq 0$, as well for an even n is proved.

Now we analyse the behaviour of Ω_1 and Ω_2 when one of the variables tends to zero or infinity, the other variable being fixed and not equal to 0 and ∞.

Let us consider $\Omega_1(z,\zeta)$. We fix $z \neq 0, \infty$. We shall prove that Ω_1 at $\zeta = 0$ has a zero, whose order is more than any pre–assigned natural number m.

We estimate the modulus Ω_1 for $|\zeta| \leq |z|$:

$$|\Omega_1(z,\zeta)| < M(r,\rho) + N(r,\rho),$$

where

$$M(r,\rho) = \frac{|\lambda|}{2\nu r^{2\nu}}\left(\frac{r}{\rho}\right)^{\nu+\frac{n}{2}}\sum_{k=n_O}^{m+n-1}\left\{H_{p_k,p_k-1}(r,\rho) + H_{p_k-1,p_k}(r,\rho)\right\},$$

$$N(r,\rho) = \frac{|\lambda|}{2\nu r^{2\nu}}\left(\frac{r}{\rho}\right)^{\nu+\frac{n}{2}}\sum_{k=m+n}^{\infty}\left\{H_{p_k,p_k-1(r,\rho)} + H_{p_k-1,p_k}(r,\rho)\right\}.$$

We also recall

$$H_{\alpha,\beta}(r,\rho) = I_\alpha\left(\frac{|\lambda|}{2\nu}r^{-2\nu}\right)K_\beta\left(\frac{|\lambda|}{2\nu}\rho^{-2\nu}\right)$$

and n_O is the nearest integer to the left of $(n+1)/2$. With regard to the asymptotic formula [26, 74]

$$K_p(x) = \sqrt{\frac{\pi}{2x}}e^{-x}\left[1 + 0(x^{-1})\right], x \to \infty,$$

we conclude that the function $M(r,\rho)$ has a zero of infinite order at $\rho = 0, r \neq 0$.

Writing $N(r,\rho)$ in the form

$$N = \frac{|\lambda|}{2\nu r^{2\nu}}\left(\frac{r}{\rho}\right)^{\nu+\frac{n}{2}} \sum_{k=m+n}^{\infty} \left\{ I_{p_k}\left[\left(\frac{\rho}{r}\right)^{2\nu}\frac{|\lambda|}{2\nu}\rho^{-2\nu}\right] K_{p_k-1}\left(\frac{|\lambda|}{2\nu}\rho^{-2\nu}\right) \right.$$
$$\left. + I_{p_k-1}\left[\left(\frac{\rho}{r}\right)^{2\nu}\frac{|\lambda|}{2\nu}\rho^{-2\nu}\right] K_{p_k}\left(\frac{|\lambda|}{2\nu}\rho^{-2\nu}\right)\right\}$$

and using (2.4) we get

$$N \leq \frac{|\lambda|}{2\nu r^{2\nu}}\left(\frac{r}{\rho}\right)^{\nu+\frac{n}{2}} \sum_{k=m+n}^{\infty} \left(\frac{\rho}{r}\right)^{2\nu(p_k-1)} \left\{ I_{p_k}\left[\frac{|\lambda|}{2\nu}\rho^{-2\nu}\right] K_{p_k-1}\left(\frac{|\lambda|}{2\nu}\rho^{-2\nu}\right) \right.$$
$$\left. + I_{p_k-1}\left(\frac{|\lambda|}{2\nu}\rho^{-2\nu}\right) K_{p_k}\left(\frac{|\lambda|}{2\nu}\rho^{-2\nu}\right)\right\}, \quad 0 \leq \rho < r.$$

Using also (2.4′), we derive the following inequality

$$N(r,\rho) \leq \sum_{k=m+n}^{\infty} \left(\frac{\rho}{r}\right)^{k-m} = \frac{\rho^m}{(r-\rho)r^{m-1}}, \quad 0 \leq \rho < r.$$

Hence it follows that the function $N(r,\rho)$ and therefore $\Omega_1(z,\zeta)$ has a zero of order m at $\zeta = 0$.

To prove relation (2.2) as $\zeta \to \infty$ (z is fixed) we estimate the modulus of Ω_1 for $|z| \leq |\zeta|$:

$$|\Omega_1(z,\zeta)| \leq M_1(\rho,r) + N_1(\rho,r),$$

where

$$M_1(\rho,r) = \frac{|\lambda|}{2\nu r^{2\nu}}\left(\frac{r}{\rho}\right)^{\nu+\frac{n}{2}} \sum_{k=n_O}^{|n_O|} \left\{ H_{p_k-1,p_k}(\rho,r) + H_{p_k,p_k-1}(\rho,r)\right\},$$

$$N_1(\rho,r) = \frac{|\lambda|}{2\nu r^{2\nu}}\left(\frac{r}{\rho}\right)^{\nu+\frac{n}{2}} \sum_{k=|n_O|+1}^{\infty} \left\{ H_{p_k-1,p_k}(\rho,r) + H_{p_k,p_k-1}(\rho,r)\right\}.$$

Having obtained an estimate for $N_1(r,\rho)$, according to the scheme we have

$$N_1(\rho,r) \leq \sum_{k=|n_O|+1}^{\infty} \left(\frac{r}{\rho}\right)^{k} = \frac{r^{|n_O|+1}}{(\rho-r)\rho^{|n_O|}}, \quad r < \rho \to \infty.$$

On the basis of the asymptotic representation [74]

$$I_p(x) \approx \frac{1}{\Gamma(p+1)}\left(\frac{x}{2}\right)^p, \quad x \to 0,$$

we get for $M_1(\rho, r)$:

$$M_1(\rho, r) = O(\rho^{-|n_O|}), \quad \rho \to \infty.$$

Thus (2.2) is proved for Ω_1.

In a similar way the behaviour of Ω_2 at $\zeta = O$ and $\zeta = \infty$ (for a fixed $z = z_O$) and the behaviour of Ω_1 and Ω_2 at $z = O$ and $z = \infty$ (for a fixed $\zeta = \zeta_0$) is established.

Lemma 2.1 is proved.

2.2 Let us consider $r^{4\nu}\Omega_1^O(z,\zeta)$ and $r^{4\nu}\Omega_2^O(z,\zeta)$. From the proof of Lemma 2.1 it follows that

$$|r^{4\nu}\Omega_1^O(z,\zeta)| < \alpha_1, \qquad |r^{4\nu}\Omega_2^O(z,\zeta)| < \alpha_2 + \alpha_3 r^4$$

where $\alpha_1, \alpha_2, \alpha_3$ are constants depending on λ, ν, n. Moreover, the functions $r^{4\nu}\Omega_1^O(z,\zeta)$ and $r^{4\nu}\Omega_2^O(z,\zeta)$ are continuous in the two variables z and ζ for $|z|, |\zeta| < R$, where R is an arbitrary large number.

The functions $\rho^{4\nu}\Omega_1^O(z,\zeta), \rho^{4\nu}\Omega_2^O(z,\zeta)$ and $(r\rho)^{4\nu}\Omega_1^O(z,\zeta), (r\rho)^{4\nu}\Omega_2^O(z,\zeta)$ possess similar properties. For the first pair of functions the inequalities are just

$$|\rho^{4\nu}\Omega_1^O(z,\zeta)| < b_1, \qquad |\rho^{4\nu}\Omega_2^O(z,\zeta)| < b_2 + b_3\,\rho^{4\nu},$$

where b_1, b_2, b_3 are constants depending on λ, ν and n.

2.3 We show the differential properties of Ω_1 and Ω_2. Differentiating term–by–term in formulas (1.13), we check immediately the relations:

$$\left\{\begin{aligned} \partial_{\bar z}\Omega_1 &= \frac{\lambda}{2r^{1+2\nu}}e^{i(n+1)\varphi}\bar\Omega_2, \\ \partial_{\bar z}\Omega_2 &= \frac{\lambda}{2r^{1+2\nu}}e^{i(n+1)\varphi}\bar\Omega_1 \end{aligned}\right. \quad \left\{\begin{aligned} \partial_{\bar\zeta}\Omega_1 &= -\frac{\bar\lambda}{2\rho^{1+2\nu}}e^{-i(n+1)\psi}\Omega_2, \\ \partial_{\zeta}\Omega_2 &= -\frac{\lambda}{2\rho^{1+2\nu}}e^{i(n+1)\psi}\Omega_1. \end{aligned}\right. \tag{2.8}$$

The first of equations (2.8) can, according to the determination (2.1), be written in the form

$$\partial_{\bar z}\Omega_1^O = \frac{\lambda}{2r^{1+2\nu}}e^{i(n+1)\varphi}\bar\Omega_2, \tag{2.9}$$

2.4 Let $\lambda \to 0$. From formula (2.3) by virtue of (1.13) and the asymptotics of modified Bessel functions we get

$$\lim_{\lambda\to 0} \Omega_1^O = 0.$$

Now from (2.1) we have

$$\lim_{\lambda\to 0} \Omega_1 = \frac{\zeta}{\zeta - z} + \Psi(z,\zeta).$$

If we introduce the additional restriction $n_O = 0$, from which $\Psi = 0$ follows, then

$$\lim_{\lambda\to 0} \Omega_1 = \frac{\zeta}{(\zeta - z)}$$

The limit of Ω_2 as $\lambda \to 0$ does not exist. Therefore we can assume that for example $\Omega_2 = 0$ if $\lambda = O$.

3 Properties of an operator

In the present section the properties of the operator S_G, given by equality (1.11), are characterized. As in the previous chapters it is not necessary to restrict here the domain G to the disk $|z| < R$. We shall assume that G is a finite multiply connected domain of a complex plane z, containing the point $z = 0$ and bounded by a finite number of simple closed Ljapunov contours Γ.

3.1 We stipulate that the function $f(z)$ belongs to the class $L_p^*(\bar{G})$, if it is represented in the form

$$f(z) = z^{4\nu} f_O(z)$$

and $f_O(z)$ is p–power integrable in $G = G + \Gamma$, i.e. $f_0(z) \in L_p(\bar{G})$.

Later on we shall denote the norms of the element f in the spaces $C(\bar{G})$ and $L_p^*(G)$ by $\|f\|_C$ and $\|f\|_{L_p^*}$; in addition $\|f\|_{L_p^*} = \|f_0\|_{L_p}$.

Lemma 3.1 *The operator S_G is completely continuous as an operator acting from $L_p^*(G), p > 2$, in $C(G)$. In addition*

$$\|S_G f\|_C \leq N R^{\mathcal{X}} \|f\|_{L_p^*}, \tag{3.1}$$

where R is the maximal distance from $z = 0$ to the domain boundary; and N and *$\mathcal{X}$ are quite concrete positive numbers.*

Proof Using (2.1) we rewrite (1.11) in the form

$$S_G f = -\frac{1}{\pi}\iint_G \frac{f(\zeta)}{\zeta - z} d\xi\, d\eta$$

$$+\frac{\lambda}{\pi}\iint_G \frac{e^{in\psi}}{\bar\zeta} \ln|1 - g(z,\zeta)|\overline{f(\zeta)}\, d\xi\, d\eta - \frac{1}{\pi}\iint_G \frac{\Psi(z,\zeta)}{\zeta} f(\zeta)\, d\xi\, d\eta$$

$$-\frac{1}{\pi}\iint_G \left[\frac{\Omega_1^O(z,\zeta)}{\zeta} f(\zeta) + \frac{\Omega_2^O(z,\zeta)}{\zeta}\overline{f(\zeta)}\right] d\xi\, d\eta,$$

where $g(z,\zeta) = z/\zeta$ for $|z| \le |\zeta|$ and $g(z,\zeta) = \zeta/z$ for $|z| \ge |\zeta|$.

As follows from [69] the first integral defines a set of functions equicontinuous in $\bar G$. It is not hard to verify that the three other integrals possess this property. In addition, for the fourth integral it is necessary to use the continuity condition of $\rho^{4\nu}\Omega_1^O$ and $\rho^{4\nu}\Omega_2^O$ in the two variables z and ζ in the topological product $G \times G$.

Now we establish the uniform boundedness of the set of functions $S_G f$. By means of the first equality of (2.1) we express $S_G f$ in the form

$$S_G f = -\frac{1}{\pi}\iint_G \frac{f(z,\zeta)}{\zeta} f(\zeta) d\xi\, d\eta - \frac{1}{\pi}\iint_G \frac{\Psi(z,\zeta)}{\zeta} f(\zeta)\, d\xi\, d\eta$$
$$-\frac{1}{\pi}\iint_G \left[\frac{\Omega_1^O(z,\zeta)}{\zeta} f(\zeta) + \frac{\Omega_2(z,\zeta)}{\zeta}\overline{f(\zeta)}\right] d\xi\, d\eta. \tag{3.2}$$

Substituting $f(\zeta) = \rho^{4\nu} f_O(\zeta)$ and employing the Holder inequality [29], we get

$$|S_G f| \le \frac{1}{\pi}\|f\|_{L_p}\left\{\left(\iint_G \rho^{4\nu q} f(\zeta)\, d\xi\, d\eta\right)^{1/q}\left(\iint_G |\zeta - z|^{-r}\, d\xi\, d\eta\right)^{1/r}\right.$$
$$+ M\left(\iint_G |\zeta|^{4(\nu-1)q^*}\, d\xi\, d\eta\right)^{1/q} + \left(\iint_G |\zeta|^{-r^*}\, d\xi\, d\eta\right)^{1/r^*}$$
$$\left.\times\left[\left(\iint_G |\rho^{4\nu}\Omega_1^O|^{p^*}\, d\xi\, d\eta\right)^{1/p^*} + \left(\iint_G |\rho^{4\nu}\Omega_2|^{p^*}\, d\xi\, d\eta\right)^{1/p^*}\right]\right\},$$

where $M = \max(|n_O|, |n_O - n - 1|)$, $1/p + 1/q + 1/r = 1$, $1/p + 1/q^* = 1$, $1/p + 1/r^* + 1/p^* = 1$, $p, q, p^* > 2$, and $1 < r, q^*, r^* < 2$. Estimates of the first four integrals through R can be found in [69], §5, ch.1. The estimates of the two others are written out here:

$$\left(\iint\limits_G |\rho^{4\nu}\Omega_1^O|^{p^*}\,\mathrm{d}\xi\,\mathrm{d}\eta\right)^{1/p^*} \leq \left(\frac{\pi}{2}\right)^{1/p^*} b_1 R^{2/p^*},$$

$$\left(\iint\limits_G |\rho^{4\nu}\Omega_2|^{p^*}\,\mathrm{d}\xi\,\mathrm{d}\eta\right)^{1/p^*} \leq (2\pi)^{1/p^*} R^{2\nu+2/p^*}$$

$$\times \sum_{k=n_O}^{\infty} \frac{|\lambda|}{k-\nu-n/2}\left\{\left[(k+2\nu-n)p^*+2\right]^{-1/p^*} + \left[(k-4\nu)p^*-2\right]^{-1/p^*}\right\}.$$

The first inequality evidently follows from the estimate for Ω_1^O; see subsection 2.2. The second one is confirmed by direct calculation. On this basis the inequality (3.1) is established. $\mathcal{X}$ is subject to the values R and v being equal to one of three numbers $4\nu+1-2/p, 4\nu-1+2/q^*$ or $1-2/p$, and N is expressed through R and the constants $|\lambda|, \nu, n, p, \ldots$ (N is bounded as $R \to 0$).

Now the assertion of Lemma 3.1 concerning the completely continuous operator S_G issues from the equicontinuity and uniform boundedness of the set of functions $S_G f$ in $C(\bar{G})$.

3.2 Let us list the function classes

$$\begin{aligned} C^*(\bar{G}) &= \{f(z) : f(z) = r^{-4v} f_O(z), f_O(z) \in C(\bar{G})\},\\ C^{**}(\bar{G}) &= \{f(z) : f(z) = r^{-2v} f_O(z), f_O(z) \in C(\bar{G})\},\\ L_P^{**}(\bar{G}) &= \{f(z) : f(z) = r^{-2v} f_O(z), f_O(z) \in L_p(\bar{G})\}. \end{aligned}$$

It turns out that the operator S_G is completely continuous from $L_p, p > 2$ in C^* and from $L_p^{**}, p > 2$, in C^{**}. The proof of these assertions relies on the properties of the kernels Ω_1^O and Ω_2^O, indicated in subsection 2.2.

3.3 Now we examine the homogeneous model equation

$$\partial_{\bar{z}}\Phi - \frac{|\lambda|}{2r^{1+2\nu}} e^{i(n+1)\varphi}\bar{\Phi} = 0, \quad z \in G. \tag{3.3}$$

Let $w(z)$ and $\Phi(z)$, the solutions of equation (1.1) and (3.3) respectively, belong to the class of functions continuous in $\bar{G} = G + \Gamma$ and continuously differentiable with respect to $\bar{z}, z$ in the domain G, except the origin. Suppose also that $f(z) \in L_p^*(\bar{G}) \cap C(G-0), p > 2$.

Theorem 3.1 *The formula*

$$w(z) = \Phi(z) + S_G f, \quad z \in G, \tag{3.4}$$

gives the general solution of equation (1.1) from $C(\bar{G}) \cap C_{\bar{z}}(G-0)$.

The proof of this theorem (with an inessential reservation) repeats the text of the proof of Lemma 3.4 from §3, ch.1

3.4 Let $\lambda = 0$. As was stated in subsection 2.4,

$$\Omega_1\Big|_{\lambda=0} = \frac{\zeta}{\zeta - z} \quad \text{and} \quad \Omega_2\Big|_{\lambda=0} = 0.$$

As a result of this, from (1.11), we get

$$S_G f\Big|_{\lambda=0} = -\frac{1}{\pi}\iint_G \frac{f(\zeta)}{\zeta - z}\, \mathrm{d}\xi\, \mathrm{d}\eta \equiv T_G f,$$

i.e. the operator constructed in this case contains the general integral operator of the theory of generalized analytic functions.

4. Relation between solutions of general and model equations

In the context of the general equation with the order of the polar singularity in the coefficient strictly greater than 1 we examine the generalized Cauchy–Riemann system of the following type:

$$\partial_{\bar{z}} w - \frac{b(z)}{2r^{1+2v}} \mathrm{e}^{\mathrm{i}(n+1)\varphi} \bar{w} = F(z), \quad z \in G. \tag{4.1}$$

As well as the conditions which have been formulated in the introduction to this chapter, we assume that

$$w(z) \in C(\bar{G}) \cap C_{\bar{z}}(G-0) \quad \text{and} \quad F(z) \in C(G-0) \cap L_p^*(\bar{G}), \quad p > 2$$

$$(L_p^*(\bar{G}) = \{f(z) : f(z) = r^{4v} f_O(z), f_O(z) \in L_p(G)\}, \|f\|_{L_p^*} = \|f_O\|_{L_p}).$$

Writing down this equation in the form

$$\partial_{\bar{w}} w - \frac{b(0)}{2r^{1+2v}} \mathrm{e}^{\mathrm{i}(n+1)\varphi} \bar{w} = F(z) + \frac{b(z) - b(0)}{2r^{1+2\nu}}\, \mathrm{e}^{\mathrm{i}(n+1)\varphi} \bar{w},$$

and employing the notation $\lambda = b(0)$ and

$$f(z) = F(z) + \frac{b(z) - b(0)}{2r^{1+2\nu}} \, \mathrm{e}^{\mathrm{i}(n+1)\varphi} \overline{w}$$

we arrive at the equation

$$\partial_{\bar{z}} w - \frac{\lambda}{2r^{1+2v}} \, \mathrm{e}^{\mathrm{i}(n+1)\varphi} \overline{w} = f(z), \quad z \in G.$$

The function $f(z)$ belongs to the class $C(G - 0) \cap L_p^*(\bar{G}), p > 2$, in the restrictions, taken for it. Expanding formula (3.4) in this case, we arrive at the general integral equation to find the function $w(z)$:

$$w(z) = \Phi(z) + S_G f + P_G \overline{w}, \tag{4.2}$$

$$P_G \overline{w} = S_G f \left(\frac{b(\zeta) - b(0)}{2\rho^{1+2\nu}} \mathrm{e}^{\mathrm{i}(n+1)\varphi} \overline{w(\zeta)} \right),$$

where $\Phi(z) \in C(\bar{G}) \cap C_{\bar{z}}(G - 0)$ is the general solution of the model equation (3.3) and the operator S_G is given by equality (1.11).

Thus, formula (4.2) states the correspondence between the sets $\{w(z)\}$ and $\{\Phi(z)\}$ of solutions from the class $C(\bar{G}) \cap C_{\bar{z}}(G - 0)$ of equations (4.1) and (3.3) so that $\Phi(z)$ is uniquely determined by the given $w(z)$ and if for some $\Phi(z)$ equation (4.2) has a solution $w(z)$ then $w(z)$ will satisfy (4.1).

Because of the properties of the operator S_G (see Lemma 3.1 and the conditions for $b(z)$ at the point $z = 0$ the operator P_G is completely continuous from $C(\bar{G})$ to itself. Taking (3.1) into account we have the estimate for the norm of P_G:

$$\|P_G\|_C \leq N R^{\chi} \left\| \frac{b(z) - b(0)}{2r^{1+6v}} \right\|_{L_p}, \quad p > 2. \tag{4.3}$$

If $\|P_G\|_C < 1$, then the next result becomes evident.

Theorem 4.1 *Formula (4.2) establishes a one-to-one correspondence between the elements of the sets of continuous solutions of the general and model equations.*

The compactness condition of the operator P_G with regard to (4.3) can be provided either for 'small measure' of the domain G (R is small) or for 'small differences' $b(z)$ from $b(0)$. If this condition is not valid then we should be satisfied only with complete continuity of the operator P_G.

The integral equation (4.2) suggests, as in the previous cases, a standard way of investigating the general equation through its simplest representative. Therefore in the following sections elements of the theory of the model equation are constructed and solutions of boundary value problems are presented. One notes that the presentation of results is generally given in a concise form.

5 Investigation of the model equation

5.1 The construction of the theory of solutions for the equation

$$\partial_{\bar{z}}\Phi - \frac{\lambda}{2r^{1+2v}}\mathrm{e}^{\mathrm{i}(n+1)\varphi}\bar{\Phi} = 0, \quad z \in G, \tag{5.1}$$

relies on the Cauchy generalized integral formula:

$$\frac{1}{2\pi\mathrm{i}}\int_{\Gamma}\frac{\Omega_1(z,\zeta)}{\zeta}\Phi(\zeta)\,\mathrm{d}\zeta - \frac{\Omega_2(z,\zeta)}{\zeta}\overline{\Phi(\zeta)\,\mathrm{d}\xi}$$

$$= \begin{cases} \Phi(z) & \text{for } z \in G, \\ \frac{1}{2}\Phi(z) & \text{for } z \in \Gamma, \\ 0 & \text{for } z \notin \bar{G} = G + \Gamma. \end{cases} \tag{5.2}$$

One assumes that $\Phi(z) \in C(\bar{G}) \cap C_{\bar{z}}(G - 0)$, the domain G is bounded by a simple closed Ljapunov contour Γ and contains the point $z = O$ in its interior. The proof of this formula repeats word for word the proof of the similar formula for solutions of the model equation with first–order polar singularity in the coefficients, see §2, ch. 2. Therefore it is not reproduced here.

5.2 Formula (5.2) allows us quite simply to state the following assertion.

Theorem 5.1 *If $\Phi_n(z), n = 1, 2, \ldots$, is a uniform convergent sequence of solutions of model equation (5.1) from the class $C(G) \cap C_{\bar{z}}(G - 0)$, then the limit function*

$$\Phi(z) = \lim_{n\to\infty}\Phi_n(z)$$

will also be a solution of (5.1) from the same class.

We can formulate another version of this assertion.

Theorem 5.2 *If $\Phi_n(z), n = 1, 2, \ldots$ is a sequence of solutions of equation (5.1) from the class $\bar{C}(G) \cap C_{\bar{z}}(G - 0)$, uniformly convergent on Γ, then $\lim\limits_{n\to\infty}\Phi_n(z)$ exists in G and defines a solution of equation (5.1) from the same class.*

5.3 Formula (5.2) allows us to obtain a representation of solutions Φ by analogy with Taylor series.

Let $G = \{z : |z| < R\}$ and $\Gamma = \{z : |z| = R\}$. Substituting $\zeta = Re^{i\psi}$ into (5.2) and employing the explicit form of the kernals Ω_1 and Ω_2 (with $|z| < |\zeta|$, see (1.13) by means of term–by–term integration we have:
for even $n = 2n_O$

$$\begin{aligned}\Phi(z) = & r^{\frac{n}{2}-\nu} \sum_{k=n_O}^{\infty} K_{p_k}\left(\frac{|\lambda|}{2\nu} r^{-2\nu}\right)(|\lambda| a_k + \lambda \bar{a}_{n-k})\, e^{ik\varphi} \\ & + K_{p_k-1}\left(\frac{|\lambda|}{2\nu} r^{-2\nu}\right)(|\lambda| a_{k-n} + \lambda \bar{a}_k) e^{i(n-k)\varphi};\end{aligned} \tag{5.3}$$

for odd $n = 2n_O - 1$

$$\begin{aligned}\Phi(z) = & r^{\frac{n}{2}-\nu} K_{\frac{1}{2}}\left(\frac{|\lambda|}{2\nu} r^{-2\nu}\right)(|\lambda| a_{n_O} + \lambda \bar{a}_{n_O})\, e^{in_O\varphi} \\ & + r^{\frac{n}{2}-\nu} \sum_{k=n_O+1}^{\infty} K_{p_k}\left(\frac{|\lambda|}{2\nu} r^{-2\nu}\right)(|\lambda| a_k + \lambda \bar{a}_{n-k})\, e^{ik\varphi} \\ & + K_{p_k-1}\left(\frac{|\lambda|}{2\nu} r^{-2\nu}\right)(|\lambda| a_{n-k} + \lambda \bar{a}_k)\, e^{i(n-k)\varphi},\end{aligned} \tag{5.4}$$

where

$$a_k = \frac{I_{p_k-1}\left(\frac{|\lambda|}{2\nu} R^{-2\nu}\right)}{4\pi\nu R^{\nu+\frac{n}{2}}} \int_O^{2\pi} \Phi(Re^{i\psi}) e^{-ik\psi}\, d\psi, \tag{5.5}$$

$$a_{n-k} = \frac{I_{p_k}\left(\frac{|\lambda|}{2\nu} R^{-2\nu}\right)}{4\pi\nu R^{\nu+\frac{n}{2}}} \int_O^{2\pi} \Phi(Re^{i\psi}) e^{-i(n-k)\psi}\, d\psi, \tag{5.5'}$$

and $k = n_O, n_O+1, \ldots$, if n is odd; and $k = n_O+1, n_O+2, \ldots$, if n is even and

$$a_{n_O} = \frac{I_{-\frac{1}{2}}\left(\frac{|\lambda|}{2\nu} R^{-2\nu}\right) + I_{\frac{1}{2}}\left(\frac{|\lambda|}{2\nu} R^{-2\nu}\right)}{8\pi\nu R^{\nu+\frac{n}{2}}} \int_O^{2\pi} \Phi(Re^{i\psi}) e^{-in_O\psi} d\psi. \tag{5.5''}$$

The meaning of the notation is described in §1.

Let the function $\Phi(z)$ be absolutely integrable on Γ. We check series (5.3) and (5.4) for absolute and uniform convergence with $z \in G$.

Let $n_O \geq O$. From (5.3) we have

$$\Phi(z) \leq r^{\frac{n}{2}-v} \sum_{k=n_O}^{\infty} \left[K_{p_k}\left(\frac{|\lambda|}{2v} r^{-2v}\right) + K_{p_k-1}\left(\frac{|\lambda|}{2v} r^{-2v}\right)\right] |\lambda|(|a_k| + |a_{n-k}|). \quad (5.6)$$

Since

$$\left| \int_O^{2\pi} \Phi(Re^{i\psi}) e^{-ik\psi} d\psi \right| < M(R),$$

then

$$|a_k| < \frac{I_{p_k-1}\left(\frac{|\lambda|}{2v} R^{-2v}\right)}{4\pi v R^{v+\frac{n}{2}}} M, \qquad |a_{n-k}| = \frac{I_{p_k}\left(\frac{|\lambda|}{2v} R^{-2v}\right)}{4\pi v R^{v+\frac{n}{2}}} M, \qquad k = n_O, n_O + 1, \ldots$$

Taking this estimate into consideration, we get from (5.6)

$$\begin{aligned}
|\Phi(z)| <& \frac{|\lambda| r^{\frac{n}{2}-\nu}}{4\pi\nu R^{\nu+\frac{n}{2}}} M \sum_{k=n_O}^{\infty} \left[K_{p_k}\left(\frac{|\lambda|}{2\nu} r^{-2\nu}\right) + K_{p_k-1}\left(\frac{|\lambda|}{2\nu} r^{-2\nu}\right)\right] \\
&\times \left[I_{p_k-1}\left(\frac{|\lambda|}{2\nu} R^{-2\nu}\right) + I_{p_k}\left(\frac{|\lambda|}{2\nu} R^{-2\nu}\right)\right] \\
\equiv& \frac{|\lambda| r^{\frac{n}{2}-\nu}}{4\pi\nu R^{\nu+\frac{n}{2}}} M \sum_{k=n_O}^{\infty} \left[K_{p_k}\left(\frac{|\lambda|}{2\nu} r^{-2\nu}\right) + K_{p_k-1}\left(\frac{|\lambda|}{2\nu} r^{-2\nu}\right)\right] \\
&\times \left[I_{p_k-1}\left(\frac{|\lambda|}{2\nu}\left(\frac{r}{R}\right)^{2\nu} r^{-2\nu}\right) + I_{p_k}\left(\frac{|\lambda|}{2\nu}\left(\frac{r}{R}\right)^{2\nu} r^{-2\nu}\right)\right].
\end{aligned}$$

Employing further the inequality

$$I_p(ax) < a^p I_p(x), \quad 0 < a < 1, \quad x > 0,$$

we have

$$\begin{aligned}
|\Phi(z)| < \frac{|\lambda| r^{\frac{n}{2}-\nu}}{4\pi\nu R^{\nu+\frac{n}{2}}} M \sum_{k=n_O}^{\infty} \left(\frac{r}{R}\right)^{2\nu(p_k-1)} & \left[K_{p_k}\left(\frac{|\lambda|}{2\nu} r^{-2\nu}\right) + K_{p_k-1}\left(\frac{|\lambda|}{2\nu} r^{-2\nu}\right)\right] \\
\times \left[I_{p_k-1}\left(\frac{|\lambda|}{2\nu} r^{-2\nu}\right) + I_{p_k}\left(\frac{|\lambda|}{2\nu} r^{-2\nu}\right)\right]&.
\end{aligned}$$

As function $K_p(t)$ increases monotonically with respect to index p and $I_p(t)$ decreases monotonically with respect to p, it is not hard to verify the correctness of the inequality

$$K_{p-1}(t) I_{p-1}(t) + K_p(t) I_p(t) < K_{p-1}(t) I_p(t) + K_p(t) I_{p-1}(t) = \frac{1}{t},$$

by which

$$|\Phi(z)| < \frac{|\lambda| r^{\frac{n}{2}-\nu}}{4\pi\nu R^{\nu+\frac{n}{2}}} M \sum_{k=n_O}^{\infty} \left(\frac{r}{R}\right)^{2\nu(p_k-1)} \left[K_{p_k}\left(\frac{|\lambda|}{2\nu} r^{-2\nu}\right) I_{p_k-1}\left(\frac{|\lambda|}{2\nu} r^{-2\nu}\right)\right.$$

$$\left. + K_{p_k-1}\left(\frac{|\lambda|}{2\nu} r^{-2\nu}\right) I_{p_k}\left(\frac{|\lambda|}{2\nu} r^{-2\nu}\right)\right]$$

$$= \frac{M r^{\frac{n}{2}-\nu}}{\pi R^{\nu+\frac{n}{2}}} \sum_{k=n_O}^{\infty} \left(\frac{r}{R}\right)^{2\nu(p_k-1)}$$

$$= \frac{M}{\pi} \sum_{k=n_O}^{\infty} \left(\frac{r}{R}\right)^{k} < N(R), \quad r < R.$$

Thus the absolute and uniform convergence of series (5.3) follows with $n_O \geq O$ and $z \in G$.

In case $n_O < O$ series (5.3) should first be divided into two sums: in the first sum the index k of summation varies from n_O to 1 and in the second one from $k = 0$ to $k = \infty$. In this case the proof of convergence is like the previous proof.

We also do not prove the uniform and absolute convergence of (5.4), becausse it does not introduce anything new.

5.4 The following statements characterize the properties of the solution of the model equation at the singular point.

Theorem 5.3 *Any solution of equation (5.1) from the class* $C(\bar{G}) \cap C_{\bar{z}}(G-0)$ *has a zero of an arbitrarily high order at* $z = O$.

This is an evident consequence the fact that $\Omega_1(z,\zeta)$ and $\Omega_2(z,\zeta)$ have a zero of an arbitrarily high order at $z = 0$; see Lemma 2.1 from §2.

Further details of this result are contained in

Theorem 5.3′ *Any solution of equation (5.1) from the class* $C(\bar{G}) \cap C_{\bar{z}}(G-0)$ *is an analytic function in the two variables* z *and* $\bar{z}$ *in* $G - 0$ *and is infinitely differentiable at* $z = 0$.

Proof Analyticity of a solution outside the singular point follows from the well-known theorems of the theory of differential equations [7]. As for infinite differentiability of solutions at $z = 0$, this property is verified by direct computation.

5.5 The Cauchy generalized integral formula describes a variety of solutions for equation (5.1) in the extended complex plane E.

Theorem 5.4 (Analogy of Liouville's theorem) *Let $\Phi(z)$ be a solution of equation (5.1) from the class $C(\bar{E}) \cap G_{\bar{z}}(E-0)$ bounded at infinity.*

If $n_O > 0$, then $\Phi(z) \equiv 0$.

If $n_O \le 0$ and n is an odd number, then there are $2(|n_O|+1)$ linearly independent real bounded solutions:

$$\begin{aligned}
\Phi_{1k}(z) &= r^{\frac{n}{2}-v}\left[K_{p_k}\left(\frac{|\lambda|}{2\nu}r^{-2\nu}\right)e^{ik\varphi} + \frac{\lambda}{|\lambda|}K_{p_k-1}\left(\frac{|\lambda|}{2\nu}r^{-2v}\right)e^{i(n-k)\varphi}\right] \\
\Phi_{2k}(z) &= ir^{\frac{n}{2}-v}\left[K_{p_k}\left(\frac{|\lambda|}{2\nu}r^{-2\nu}\right)e^{ik\varphi} - \frac{|\lambda|}{|\lambda|}K_{p_k-1}\left(\frac{|\lambda|}{2\nu}r^{-2v}\right)e^{i(n-k)\varphi}\right] \\
&(k = n_O, n_O+1, \ldots, 0)
\end{aligned} \tag{5.7}$$

and

$$\Phi_{1k}(z), \quad \Phi_{2k}(z) = O(|z|^k), \qquad z \to \infty. \tag{5.8}$$

If $n_O \le 0$ and n is an even number, then there are $2|n_O|+1$ linearly independent real bounded solutions. For the index values $k = n_O+1, n_O+2, \ldots, 0$ these solutions are given by formulas (5.7); one more solution is defined by the relation

$$\Phi_{n_O}(z) = \delta(\zeta)r^{\frac{n}{2}-v}K_{\frac{1}{2}}\left(\frac{|\lambda|}{2v}r^{-2v}\right)e^{in_O\varphi}, \tag{5.9}$$

in which

$$\delta(\lambda) = \begin{cases} 1 & \text{if } \lambda > 0; \\ i & \text{if } \lambda < 0; \\ \frac{\lambda+|\lambda|}{\operatorname{Im}\lambda} & \text{if } \operatorname{Im}\lambda \ne 0 \end{cases}.$$

The behaviour of these functions as $z \to 0$ is characterized by (5.8).

Proof Since $|\Phi(z)| < M, z \in E$, then from (5.5), (5.5′) and (5.5″) it follows that

$$|a_k| < \frac{I_{p_k-1}\left(\frac{|\lambda|}{2v}R^{-2v}\right)}{2\pi\nu R^{\nu+\frac{n}{2}}}M,$$

$$|a_{n-k}| < \frac{I_{p_k}\left(\frac{|\lambda|}{2v}R^{-2v}\right)}{2\pi\nu R^{\nu+\frac{n}{2}}}M, \quad k = n_O, n_O+1, \ldots, 0$$

$$|a_{n_O}| < \frac{I_{-\frac{1}{2}}\left(\frac{|\lambda|}{2v}R^{-2v}\right) + I_{\frac{1}{2}}\left(\frac{|\lambda|}{2v}R^{-2v}\right)}{4\pi\nu R^{v+\frac{n}{2}}}M.$$

These inequalities hold for any values of R. Therefore with regard to the asymptotic formula as $R \to \infty$, see [26],

$$I_p(t) \approx \frac{1}{\Gamma(p+1)} \left(\frac{t}{2}\right)^p, \quad t \to 0,$$

we get for $n_O > 0$

$$|a_k| = |a_{n-k}| = 0,$$

i.e.

$$a_k = a_{n-k} = 0 \quad (k = n_O, n_O + 1, \dots, 0)$$

In this case we get $\Phi(z) \equiv O$ from (5.3) and (5.4).

Now let $n_O \leq 0$. By the asymptotic formula

$$K_p(t) \approx \frac{1}{\Gamma(p+1)} \left(\frac{t}{2}\right)^p, \quad t \to 0,$$

it is directly established that only (5.7), (5.9) are solutions of equation (5.1), which satisfy (5.8).

Theorem 5.3 is proved.

This theorem is generalized in the case when the solutions allow a concrete growth at infinity.

Theorem 5.5 (Analogy Liouville's generalized theorem) *Let $\Phi(z)$ be a solution of equation (5.1) from the class $C(\bar{E}) \cap C_{\bar{z}}(E-0)$ and $\Phi(z) = O(|z|^N), z \to \infty$, where N is a non-negative integer.*

If $n_O > N$, then $\Phi(z) \equiv O$.

If $n_O \leq N$, then the non-trivial variety of solutions is described by the formula

$$\begin{aligned}\Phi(z) =& r^{\frac{n}{2}-\nu} \sum_{k=n_O+1}^{N} B_k K_{p_k}\left(\frac{|\lambda|}{2\nu} r^{-2\nu}\right) e^{ik\varphi} \\ &+ \frac{\lambda}{|\lambda|} \bar{B}_k K_{p_k-1}\left(\frac{|\lambda|}{2\nu} r^{-2\nu}\right) e^{i(n-k)\varphi},\end{aligned} \tag{5.10}$$

when n is an odd number, and is expressed by the formula

$$\begin{aligned}\Phi(z) =& r^{\frac{n}{2}-\nu} \left[B_{n_O} K_{\frac{1}{2}}\left(\frac{|\lambda|}{2\nu} r^{-2\nu}\right) e^{in_O\varphi} + \sum_{k=n_O+1}^{N} B_k K_{p_k}\left(\frac{|\lambda|}{2\nu} r^{-2\nu}\right) e^{ik\varphi} \right. \\ &\left. + \frac{\lambda}{|\lambda|} \bar{B}_k K_{p_k-1}\left(\frac{|\lambda|}{2\nu} r^{-2\nu}\right) e^{i(n-k)\varphi} \right].\end{aligned} \tag{5.11}$$

when n is an even number; B_k are arbitrary complex constants and $|\lambda| B_{n_O} = \lambda \bar{B}_{n_O}$ for even numbers n.

5.6 The Cauchy generalized formula is obviously valid for the circular ring $C = \{z : R_O < |z| < R\}$. In this case we can write it in the form

$$\begin{aligned}\Phi(z) =& \frac{1}{2\pi \mathrm{i}} \int_{\Gamma} \frac{\Omega_1(z,\zeta)}{\zeta} \Phi(\zeta)\,\mathrm{d}\zeta - \frac{\Omega_2(z,\zeta)}{\bar{\zeta}} \overline{\Phi(\zeta)\,\mathrm{d}\xi} \\ & - \frac{1}{2\pi \mathrm{i}} \int_{\Gamma_O} \frac{\Omega_1(z,\zeta)}{\zeta} \Phi(\zeta)\,\mathrm{d}\zeta - \frac{\Omega_2(z,\zeta)}{\bar{\zeta}} \overline{\Phi(\zeta)\,\mathrm{d}\xi},\end{aligned} \tag{5.12}$$

where $\Gamma = \{\zeta : |\zeta| = R\}$ and $\Gamma_O = \{\zeta : |\zeta| = R_O\}$. Since on Γ_O and Γ we have $\zeta_O = R_O \mathrm{e}^{\mathrm{i}\psi_o}$ and $\zeta = R\mathrm{e}^{\mathrm{i}\psi}$ respectively, then assuming $z \in G$ and introducing Ω_1, Ω_2 from (1.13) in (5.12), via term–by–term integration we get
for odd $n = 2n_O - 1$

$$\begin{aligned}\Phi(z) =& r^{\frac{n}{2}-\nu} \Bigg[\sum_{k=n_O}^{\infty} K_{p_k}\left(\frac{|\lambda|}{2\nu} r^{-2\nu}\right) (|\lambda| a_k + \lambda \overline{a_{n-k}})\, \mathrm{e}^{\mathrm{i}k\varphi} \\ & + K_{p_k - 1}\left(\frac{|\lambda|}{2\nu} r^{-2\nu}\right) (|\lambda| a_{n-k} + \lambda \overline{a_k})\, \mathrm{e}^{\mathrm{i}(n-k)\varphi} \\ & + \sum_{k=n_O}^{\infty} I_{p_k}\left(\frac{|\lambda|}{2\nu} r^{-2\nu}\right) (|\lambda| b_k - \lambda \overline{b_{n-k}})\, \mathrm{e}^{\mathrm{i}k\varphi} \\ & + I_{p_k - 1}\left(\frac{|\lambda|}{2\nu} r^{-2\nu}\right) (|\lambda| b_{n-k} - \lambda \overline{b_k})\, \mathrm{e}^{\mathrm{i}(n-k)\varphi} \Bigg];\end{aligned} \tag{5.13}$$

for even n

$$\begin{aligned}\Phi(z) =& r^{\frac{n}{2}-\nu} \Bigg[K_{\frac{1}{2}}\left(\frac{|\lambda|}{2\nu} r^{-2\nu}\right) (|\lambda| a_{n_O} + \lambda \overline{a_{n_O}})\, \mathrm{e}^{\mathrm{i}n_O\varphi} \\ & + \sum_{k=n_O+1}^{\infty} K_{p_k}\left(\frac{|\lambda|}{2\nu} r^{-2\nu}\right) (|\lambda| a_k + \lambda \overline{a_{n-k}})\, \mathrm{e}^{\mathrm{i}k\varphi} \\ & + \frac{1}{2}\left\{ I_{\frac{1}{2}}\left(\frac{|\lambda|}{2\nu} r^{-2\nu}\right) + I_{-\frac{1}{2}}\left(\frac{|\lambda|}{2\nu} r^{-2\nu}\right) \right\} (|\lambda| b_{n_O} - \lambda \overline{b_{n_O}})\, \mathrm{e}^{\mathrm{i}n_O\varphi} \\ & + \sum_{k=n_O+1}^{\infty} I_{p_k}\left(\frac{|\lambda|}{2\nu} r^{-2\nu}\right) (|\lambda| b_k - \lambda \overline{b_{n-k}})\, \mathrm{e}^{\mathrm{i}k\varphi} \\ & + I_{p_k - 1}\left(\frac{|\lambda|}{2\nu} r^{-2\nu}\right) (|\lambda| b_{n-k} - \lambda \overline{b_k})\, \mathrm{e}^{\mathrm{i}(n-k)\varphi}\end{aligned} \tag{5.14}$$

where

$$
\begin{aligned}
b_k &= \frac{K_{p_k-1}\left(\frac{|\lambda|}{2\nu}R_O^{-2\nu}\right)}{4\pi\nu R_O^{\nu+\frac{n}{2}}} \int_O^{2\pi} \Phi(R_O e^{i\psi_O}) e^{-ik\psi_O}\, d\psi_O, \\
b_{n-k} &= \frac{K_{p_k}\left(\frac{|\lambda|}{2\nu}R_O^{-2\nu}\right)}{4\pi\nu R_O^{\nu+\frac{n}{2}}} \int_O^{2\pi} \Phi(R_O e^{i\psi_O}) e^{-i(n-k)\psi_O}\, d\psi_O, \\
&(k = n_O, n_O+1, \ldots)
\end{aligned}
\tag{5.15}
$$

and the expressions for a_{n_o}, a_k, a_{n-k} are those described in formulas (5.5), (5.5′), (5.5″).

If we suppose that $\Phi(z)$ is absolutely integrable on Γ_O and Γ in the variables ψ_O and ψ respectively, then it will be sufficient for the absolute and uniform convergence of the first series in (5.13) with $|z| > R_O$. As to the simultaneous convergence of both series, it takes place in the circular ring $R_O < |z| < R$. It is well known that this refers to (5.14) as well.

By means of (5.13) and (5.14) the classification of singularities $\Phi(z)$ at the point $z = 0$ can be done. Without dwelling on it, we state that using the above formulas it is not difficult to prove a generalization of Theorem 5.5 based on an asymptotic representation for modified Bessel functions.

Theorem 5.6 *Let $\Phi(z)$ be a solution of equation (5.1) from the class $C(E-0) \cap C_{\bar{z}}(E-0)$ and*

$$
|\Phi(z)| < \frac{M}{|z|^l}, \quad z \to 0,
$$
$$
\Phi(z) = O(|z|^N), \quad z \to \infty,
$$

where l is some positive number and N is a natural number.

If $n_O > N$, then $\Phi(z) \equiv 0$.

If $n_O \leq N$, then the variety of solutions is given by the formulas

$$
\Phi(z) = r^{\frac{n}{2}-v} \sum_{k=n_O}^{N} B_k K_{p_k}\left(\frac{|\lambda|}{2\nu} r^{-2\nu}\right) e^{ik\varphi} + \frac{\lambda}{|\lambda|}\bar{B}_k K_{p_k-1}\left(\frac{|\lambda|}{2\nu} r^{-2\nu}\right) e^{i(n-k)\varphi},
$$

(for odd n),

$$
\Phi(z) = r^{\frac{n}{2}-v}\left[B_{n_O} K_{\frac{1}{2}}\left(\frac{|\lambda|}{2\nu} r^{-2\nu}\right) e^{in_O\varphi} + \sum_{k=n_O+1}^{N} B_k K_{p_k}\left(\frac{|\lambda|}{2\nu} r^{-2\nu}\right) e^{ik\varphi}\right.
$$

(for even n).

$$+\frac{\lambda}{|\lambda|}\bar{B}_k K_{p_k-1}\left(\frac{|\lambda|}{2\nu}r^{-2\nu}\right)e^{i(n-k)\varphi}\Bigg].$$

In both formulas B_k are arbitrary complex constants and $|\lambda|B_{n_O} = \lambda\bar{B}_{n_O}$ for even n.

5.7 Let G^+ be a simply connected domain, bounded by a simple closed Ljapunov contour Γ. Let us denote the supplement of the domain $G^+ + \Gamma$ to the whole widened complex plane E by G^-. We examine the integral

$$\Phi(z) = \frac{1}{2\pi i}\int_\Gamma \frac{\Omega_1(z,\zeta)}{\zeta}\mu(\zeta)\,d\zeta - \frac{\Omega_2(z,\zeta)}{\bar{\zeta}}\overline{\mu(\zeta)\,d\xi}, \tag{5.16}$$

where $\mu(\zeta)$ is a function satisfying the Holder condition on Γ, and Ω_1 and Ω_2 are the general kernels given by equality (1.13).

This integral defines the functions $\Phi^+(z)$ and $\Phi^-(z)$ in the domains G^+ and G^-. Either of them is a solution of equation (5.1) and $\Phi^+(z) \in C(G^+ + \Gamma) \cap C_{\bar{z}}(G^+ - 0)$ and $\Phi^-(z) \in C(G^+ + \Gamma - \infty) \cap C^1(G^- - \infty)$.

The asymptotic behaviour of $\Phi^-(z)$ for $z \to \infty$ is characterized by the properties of Ω_1 and Ω_2, see (2.2′) from §2:

$$\Phi^-(z) = O(|z|^{n-n_O}), \quad z \to \infty.$$

It is evident that this function has a zero of order $|n - n_O|$ for $n_O \le 0$ and a pole of $n - n_O$ for $n_O > 0$.

Formula (5.16), generalizing the integral of Cauchy type establishes additional properties of the solution to equation (5.1). In particular, if we assume $z = t \in \Gamma$, then (5.16) could be written in the form

$$\Phi(t) = \frac{1}{2\pi i}\int_\Gamma \frac{\Omega_1(t,\zeta)}{\zeta}\mu(\zeta)\,d\zeta - \frac{\Omega_2(t,\zeta)}{\bar{\zeta}}\overline{\mu(\zeta)\,d\xi}.$$

This integral exists in the sense of the Cauchy principal value and defines a function $\Phi(t)$ satisfying the Holder condition on Γ.

If we suppose that $z \in G^+$ (or G^-) in (5.16), and pass over to the limit as $z \to t \in \Gamma$, then we come to an analogy of the well-known Sohockiii–Plemelj formulas

$$\Phi^+(t) = \frac{1}{2}\mu(t) + \Phi(t),$$

$$\Phi^-(t) = -\frac{1}{2}\mu(t) + \Phi(t), \quad t \in \Gamma,$$

where $\Phi(t)$ is given by formula (5.16). Combining them we get their equivalent form:

$$\Phi^+(t) - \Phi^-(t) = \mu(t), \qquad \Phi^+(t) + \Phi^-(t) = \Phi(t).$$

5.8 Using the generalized integral of Cauchy type, we can solve the problem of searching for a piecewise regular solution $\Phi(z) = \Phi^{\pm}(z), z \in G^{\pm}$, of equation (5.1), bounded at $z = \infty$ and satisfying the equality

$$\Phi^+(t) - \Phi^-(t) = \mu(t), \quad t \in \Gamma, \tag{5.17}$$

where $\mu(t)$ is a given function subordinated to the Holder condition.

In fact, it was previously stated that the function

$$\Phi^*(z) = \frac{1}{2\pi i} \int_\Gamma \frac{\Omega_1(z,\zeta)}{\zeta}\mu(\zeta)\,d\zeta - \frac{\Omega_2(z,\zeta)}{\bar{\zeta}}\overline{\mu(\zeta)\,d\xi}, \tag{5.18}$$

satisfies equality (5.17). For $n_O \leq 0$ it has a zero of order $|n - n_O|$. Therefore $\Phi^*(z)$ is a partial solution of this problem.

Let us change the unknown function $\Phi(z)$ by the formula

$$\Phi(z) = \Phi_O(z) + \Phi^*(z).$$

The new unknown function $\Phi_O(z)$ is a solution of equation (5.1), bounded at infinity and thanks to the condition

$$\Phi_O^+(t) = \Phi_O^-(t) \quad \text{on } \Gamma$$

is continuous on the plane E. Applying Theorem 5.4 (the analogy of Liouville's theorem) to our search we get the complete description of the set of unknown functions $\Phi(z)$:

$$\Phi(z) = \Phi^*(z) + \sum_{k=n_O}^{n} a_{1k}\Phi_{1k}(z) + a_{2k}\Phi_{2k}(z), \quad (\text{for odd } n) \tag{5.19}$$

and

$$\Phi(z) = \Phi^*(z) + a_O\Phi_{n_o}(z) + \sum_{k=n_O+1}^{0} a_{1k}\Phi_{1k}(z) + a_{2k}\Phi_{2k}(z), \quad (\text{for even } n). \tag{5.20}$$

Here a_O, a_{1k}, a_{2k} are arbitrary real constants, $\Phi^*(z)$ is defined by formula (5.18), and $\Phi_{n_O}, \Phi_{1k}, \Phi_{2k}$ are defined by functions (5.7), (5.9).

Now we examine the case $n_O > 0$. The function $\Phi^*(z)$, given by relation (5.18), satisfies condition (5.17), but as has been noted in subsection 5.7, it has a pole of order $n - n_O$ at the point $z = 0$. And if we again exchange the unknown function $\Phi(z)$ with a new function $\Phi_O(z)$ by means of $\Phi^*(z)$, then $\Phi_O(z)$ will turn out to be continuous on the complex plane and unbounded at infinity, where it will have a pole of the same order as $\Phi^*(z)$.

We apply the results of Theorem 5.5 (the analogy of Liouville's generalized theorem) to find $\Phi_O(z)$.

Let $n = 2n_O - 1$ be an odd number. As $N = n - n_O = n_O - 1 < n_O$, then $\Phi_O(z) \equiv 0$ and $\Phi(z) = \Phi^*(z)$. To prove the boundedness of $\Phi(z)$ as $z \to \infty$ is necessary and sufficient to impose the conditions:

$$\int_\Gamma \rho^{-\nu-\frac{n}{2}-1} K_{p_k-1}\left(\frac{|\lambda|}{2\nu}\rho^{-2\nu}\right) e^{-i(k+1)\psi}\mu(\zeta)\,d\zeta = 0 \tag{5.21}$$

$$(k = 0, 1, \dots, n).$$

To obtain these conditions we should introduce explicit expressions of Ω_1 and Ω_2 in formula (5.13), writing down the representation of $\Phi^*(z)$ in series form in a neighbourhood of $z = \infty$ and lastly equating the first $(n+1)$ coefficients of the series to zero because they are unbounded at infinity.

Now let $n = 2n_O$ be an even number. In this case the previous procedure can be repeated. The difference is in the fact that $N = n - n_O = n_O$ and therefore from Theorem 5.5 it follows that

$$\Phi_O(z) = r^{\frac{n}{2}-\nu} B_{n_O} K_{\frac{1}{2}}\left(\frac{|\lambda|}{2\nu} r^{-2\nu}\right) e^{in_O\varphi},$$

where the complex constant B_{n_O} must satisfy the relation $|\lambda| B = \lambda \bar{B}_{n_O}$.

For $\Phi(z)$ we get

$$\begin{aligned}\Phi(z) =& \Phi_O(z) + \Phi^*(z) = r^{\frac{n}{2}-\nu} B_{n_O} K_{\frac{1}{2}}\left(\frac{|\lambda|}{2\nu} r^{-2\nu}\right) e^{in_O\varphi} \\ &+ \frac{1}{2\pi i}\int_\Gamma \frac{\Omega_1(z,\zeta)}{\zeta}\mu(\zeta)\,d\zeta - \frac{\Omega_2(z,\zeta)}{\bar{\zeta}}\overline{\mu(\zeta)\,d\xi}.\end{aligned}$$

and $\Phi(z)$ is not bounded as $z \to \infty$. Again expanding $\Phi^*(z)$ in a series in a neighbourhood of $z = \infty$ and equating to zero the coefficients of those summands of the series which have a pole, we come to conditions (5.21) with $k = 0, 1, \ldots, n_O - 1, n_O + 1, \ldots, n$ but for $k = n_O$

$$\frac{|\lambda|}{8\nu\pi \mathrm{i}} A - \frac{\pi}{2} B_{n_O} = 0, \qquad \frac{|\lambda|}{8\nu\pi \mathrm{i}} A + \frac{\pi}{2} B_{n_O} = 0,$$

where

$$A = \int_\Gamma \rho^{-\nu - n_O - 1} K_{\frac{1}{2}} \left(\frac{|\lambda|}{2\nu} \rho^{-2\nu} \right) \left[\mathrm{e}^{-\mathrm{i}(n_O+1)\psi} \mu(\zeta)\,\mathrm{d}\zeta + \mathrm{e}^{\mathrm{i}(n_O+1)\psi} \overline{\mu(\zeta)\,\mathrm{d}\xi} \right].$$

Hence it follows $B_{n_O} = 0$ and

$$A = 0. \tag{5.22}$$

Thus we have established

Theorem 5.7 *Let* $n_O \leq 0$.

If $n = 2n_O - 1$, *then the problem of discovering a piecewise regular solution* $\Phi(z)$ *of equation (5.1), bounded at the point* $z = \infty$ *and satisfying condition (5.17), has* $2(|n_O| + 1)$ *linearly independent solutions, given by formulas (5.7). The general solution contains* $2(|n_O| + 1)$ *arbitrary real constants and is determined by formula (5.19).*

If $n = 2n_O$, *then the number of linearly independent solutions is equal to* $2|n_O| + 1$, *and the general solution contains* $2|n_O| + 1$ *arbitrary real constants and is given by relation (5.20).*

Let $n_O > 0$. *In this case the problem is generally speaking unsolvable. If* $\mu(t), t \in \Gamma$, *satisfies condition (5.21) for an odd* n *and conditions (5.21), (5.22) for an even* n, *then the unique solution of the problem is given by formula (5.18).*

6 Boundary value problems for solutions of the model equation

In the present section the results of analysing the boundary value problems of Dirichlet and Riemann–Hilbert and the linear conjugation problem for solutions $\Phi(z)$ of the equation

$$\partial_{\bar{z}} \Phi - \frac{\lambda}{2r^{1+2\nu}} \mathrm{e}^{\mathrm{i}(n+1)\varphi} \bar{\Phi} = 0. \tag{6.1}$$

are presented. The unknown solution is sought in a circular domain $G = \{z : |z| < R\}$, and boundary problems are given in a canonical form on the circle $\Gamma = \{z : |z| = R\}$.

In this formulation the solution of the boundary value problems will be obtained in explicit form, using the method of expanding a solution in Fourier series .

6.1 Dirichlet problem *It is necessary to find in $G = \{z : |z| < R\}$ a solution $\Phi(z)$ of equation (6.1), subordinated on the boundary $\Gamma = \{z : |z| = Re^{i\varphi}\}$ to the condition:*

$$\text{Re } \Phi(z) = h(z), \quad z \in \Gamma. \tag{6.2}$$

One assumes that $\Phi(z) \in C(\bar{G}) \cap C_{\bar{z}}(G - 0)$ and

$$h(z) = h(Re^{i\varphi}) = h_O + \frac{1}{2}\sum_{k=1}^{\infty} h_k e^{ik\varphi} + \bar{h}_k e^{-ik\varphi}.$$

To investigate this problem we use a representation of the solution in analogy to Taylor series, see (5.3) and (5.4):
for an odd n:

$$\begin{aligned}\Phi(z) =& r^{\frac{n}{2}-\nu} \sum_{k=n_O}^{\infty} B_k K_{p_k}\left(\frac{|\lambda|}{2\nu} r^{-2\nu}\right) e^{ik\varphi} \\ &+ \frac{\lambda}{|\lambda|} \bar{B}_k K_{p_k - 1}\left(\frac{|\lambda|}{2\nu} r^{-2\nu}\right) e^{i(n-k)\varphi},\end{aligned} \tag{6.3}$$

for an even n:

$$\begin{aligned}\Phi(z) =& r^{\frac{n}{2}-v} B_{n_O} K_{\frac{1}{2}}\left(\frac{|\lambda|}{2\nu} r^{-2\nu}\right) e^{in_O\varphi} + r^{\frac{n}{2}-\nu} \sum_{k=n_o+1}^{N} B_k K_{p_k}\left(\frac{|\lambda|}{2\nu} r^{-2\nu}\right) e^{ik\varphi} \\ &+ \frac{\lambda}{|\lambda|} \bar{B}_k K_{p_k - 1}\left(\frac{|\lambda|}{2\nu} r^{-2\nu}\right) e^{i(n-k)\varphi},\end{aligned} \tag{6.4}$$

where B_k are arbitrary complex constants and $|\lambda| B_{n_O} = \lambda \bar{B}_{n_O}$ for even n, n_O is the nearest integer to the left of $(n+1)/2$ and $p_k = (k + \nu - \frac{n}{2})/2\nu$.

Introducing (6.3) and (6.4) into the boundary problem (6.2) and equating coefficients of the same power of $e^{i\varphi}$ we discover that B_k must be subordinated to the following algebraic equations:
with $n \leq 0$

$$(B_O + \bar{B}_O) K_{p_O}\left(\frac{|\lambda|}{2\nu} R^{-2\nu}\right) = 2h_O R^{\nu - \frac{n}{2}}, \tag{6.5}$$

$$B_s K_{p_s}\left(\frac{|\lambda|}{2\nu} R^{-2\nu}\right) + \bar{B}_{-s} K_{p_{-s}}\left(\frac{|\lambda|}{2\nu} R^{-2\nu}\right) = h_s R^{\nu - \frac{n}{2}}, \tag{6.6}$$

$$(s = 1, 2, \ldots \left[\frac{|n|}{2}\right]),$$

$$B_s K_{p_s}\left(\frac{|\lambda|}{2\nu}R^{-2\nu}\right) + \frac{\bar{\lambda}}{|\lambda|} B_{n+s} K_{p_{-s}}\left(\frac{|\lambda|}{2\nu}R^{-2\nu}\right) = h_s R^{\nu-\frac{n}{2}} \tag{6.7}$$

$$\left(s = \left[\frac{|n|}{2}\right] + 1, \ldots\right);$$

with $n > 0$;

$$\left(\frac{\lambda}{|\lambda|}\bar{B}_n + \frac{\bar{\lambda}}{|\lambda|} B_n\right) K_{pO}\left(\frac{|\lambda|}{2\nu}R^{-2\nu}\right) = 2h_O R^{\nu-\frac{n}{2}}, \tag{6.8}$$

$$\frac{\lambda}{|\lambda|}\bar{B}_{n-s} K_{p_s}\left(\frac{|\lambda|}{2\nu}R^{-2\nu}\right) + \frac{\bar{\lambda}}{|\lambda|} B_{b+s} K_{p_{-s}}\left(\frac{|\lambda|}{2\nu}R^{-2\nu}\right) = h_s R^{\nu-\frac{n}{2}} \tag{6.9}$$

$$\left(s = 1, 2, \ldots, \left[\frac{|n|}{2}\right]\right),$$

$$B_s K_{p_s}\left(\frac{|\lambda|}{2\nu}R^{-2\nu}\right) + \frac{\bar{\lambda}}{|\lambda|} B_{n+s} K_{p_{-s}}\left(\frac{|\lambda|}{2\nu}R^{-2\nu}\right) = h_s R^{\nu-\frac{n}{2}} \tag{6.10}$$

$$\left(s = \left[\frac{|n|}{2}\right] + 1, \ldots\right).$$

Later it will be necessary to examine certain cases $n \leq -1, n = 0$ and $n \geq 1$. As before the question concerns only real solvability conditions and the linearly independence over the field of real numbers. The boundary problem is called homogeneous if $h(Re^{i\varphi}) = 0$.

Theorem 6.1 *Let $n \leq -1$. Then the homogeneous Dirichlet problem has $|n|$ linearly independent solutions and the inhomogeneous problem is undoubtedly solvable.*

Proof Let n be odd. First we examine the homogeneous problem. It generates an infinite system of linear algebraic equations which are extracted from (6.5)–(6.7) with $h_s = 0, s = 0, 1, \ldots$; these will be denoted by $(6.5^\circ) - (6.7^\circ)$.

From (6.5) it follows that $B_O = ic_O$, where c_O is an arbitrary real constant. The coefficients $B_{-nk} (k = 1, 2, \ldots)$ are defined by the following recurrence relations through B_O:

$$B_{-nk} K_{p_{-nk}}\left(\frac{|\lambda|}{2\nu}R^{-2\nu}\right) + \frac{\bar{\lambda}}{|\lambda|}\bar{B}_{n(1-k)} K_{p_{nk}}\left(\frac{|\lambda|}{2\nu}R^{-2\nu}\right) = 0,$$

which are obtained from (6.7°). Hence we have

$$B_{-nk} = \left(-\frac{\bar{\lambda}}{|\lambda|}\right)^k \prod_{m=1}^{k} \frac{K_{p_{-mn}-1}\left(\frac{|\lambda|}{2\nu}R^{-2\nu}\right)}{K_{p_{-mn}}\left(\frac{|\lambda|}{2\nu}R^{-2\nu}\right)} \cdot ic_O \quad (k = 1, 2, \ldots) \tag{6.11}$$

where $\prod$ is the product symbol.

The real constant c_O generates one non–trivial solution

$$\begin{aligned}\Phi_O(z) = \sum_{k=0}^{\infty} B_{-nk} K_{p_{-nk}}\left(\frac{|\lambda|}{2\nu} r^{-2\nu}\right) e^{-ink\varphi} \\ + \frac{\lambda}{|\lambda|} \bar{B}_{-nk} K_{p_{-nk}-1}\left(\frac{|\lambda|}{2\nu} r^{-2\nu}\right) e^{in(k+1)\varphi},\end{aligned} \tag{6.12}$$

of the homogeneous problem, B_{-nk} being expressed by c_O; see (6.11).

The series (6.9) absolutely and uniformly converges in the closed circle $|z| \leq R$. In fact, as $K_p(t)$ is monotonically decreasing with respect to t, then

$$|\Phi_O(z)| < \sum_{k=0}^{\infty} |B_{-nk}| K_{p_{-nk}}\left(\frac{|\lambda|}{2\nu} R^{-2\nu}\right) + \sum_{k=0}^{\infty} |B_{-nk}| K_{p_{-nk}-1}\left(\frac{|\lambda|}{2\nu} R^{-2\nu}\right). \tag{6.13}$$

The first series in the right part of (6.13) converges on account of d'Alambert's test and the relation

$$\lim_{k\to\infty} \left|\frac{B_{-n(k+1)}}{B_{-nk}}\right| \frac{K_{p_{-n(k+1)}}\left(\frac{|\lambda|}{2\nu} R^{-2\nu}\right)}{K_{p_{-nk}}\left(\frac{|\lambda|}{2\nu} R^{-2\nu}\right)}$$

$$= \lim_{k\to\infty} \frac{K_{p_{-nk}-1}\left(\frac{|\lambda|}{2\nu} R^{-2\nu}\right)}{K_{p_{-nk}}\left(\frac{|\lambda|}{2\nu} R^{-2\nu}\right)} = q < 1.$$

Here we have used the expression for B_{-nk} through c_O and the monotone increase of $K_p(t)$ with respect to p. In exactly the same way the convergence of the second series in (6.13) is established.

Now we turn to equation (6.6°). If we suppose $B_{-q} = c_{-q} (q = 1, 2, \ldots, \left[\frac{|n|}{2}\right])$, where c_{-q} is an arbitrary complex number, then the following relation issues from (6.6°) :

$$B_q = -\frac{K_{p_{-q}}\left(\frac{|\lambda|}{2\nu} R^{-2\nu}\right)}{K_{p_q}\left(\frac{|\lambda|}{2\nu} R^{-2\nu}\right)} \bar{c}_{-q},$$

and with help of the sequence of formulas

$$B_{-nk\mp q}K_{p_{-nk\mp q}}\left(\frac{|\lambda|}{2\nu}R^{-2\nu}\right)+\frac{\bar{\lambda}}{|\lambda|}B_{-n(k+1)\mp q}K_{p_{nk\mp q}}\left(\frac{|\lambda|}{2\nu}R^{-2\nu}\right)=0,$$

$k=1,2,\ldots$, all the coefficients of the type $B_{-nk\mp q}(k=1,2,\ldots)$ are defined:

$$B_{-nk+q}=\left(-\frac{\bar{\lambda}}{|\lambda|}\right)^{k}\prod_{m=O}^{k}\frac{K_{p_{nm-q}}\left(\frac{|\lambda|}{2\nu}R^{-2\nu}\right)}{K_{p_{-nm+q}}\left(\frac{|\lambda|}{2\nu}R^{-2\nu}\right)}\bar{c}_{-q},$$

$$B_{-nk-q}=\left(-\frac{\bar{\lambda}}{|\lambda|}\right)^{k}\prod_{m=O}^{k}\frac{K_{p_{nm+q}}\left(\frac{|\lambda|}{2\nu}R^{-2\nu}\right)}{K_{p_{-nm-q}}\left(\frac{|\lambda|}{2\nu}R^{-2\nu}\right)}\bar{c}_{-q},\tag{6.14}$$

The arbitrary constants $c_{-q}(q=1,2,\ldots,\left[\frac{|n|}{2}\right])$ generate non–trivial solutions of the homogeneous problem (6.1), (6.2):

$$\begin{aligned}\Phi_{-q}(z)=&r^{\frac{n}{2}-\nu}\sum_{k=0}^{\infty}\Big\{B_{-nk+q}K_{p_{-nk+q}}\left(\frac{|\lambda|}{2\nu}r^{-2\nu}\right)e^{i(-nk+q)\varphi}\\&+\frac{\lambda}{|\lambda|}\bar{B}_{-nk+q}K_{p_{-nk+q}-1}\left(\frac{|\lambda|}{2\nu}r^{-2\nu}\right)e^{i[n(k+1)-q]\varphi}\\&+B_{-nk-q}K_{p_{-nk-q}}\left(\frac{|\lambda|}{2\nu}r^{-2\nu}\right)e^{-i(nk+q)\varphi}\\&+\frac{\lambda}{|\lambda|}\bar{B}_{-nk-q}K_{p_{-nk-q}-1}\left(\frac{|\lambda|}{2\nu}r^{-2\nu}\right)e^{i[n(k+1)+q]\varphi}\Big\}\end{aligned}\tag{6.15}$$

where the constant $B_{-nk\mp q}$ is expressed by c_{-q} from formula (6.14). The absolute and uniform convergence of series (6.15) is checked in the same way as (6.12).

Thus, the general solution $\Phi^*(z)$ of the homogeneous problem has the form:

$$\Phi^*(z)=\Phi_O(z)+\sum_{q=1}^{\left[\frac{|n|}{2}\right]}\Phi_{-q}(z).\tag{6.16}$$

Introducing the notation

$$\hat{\Phi}_O(z)=[\Phi_O(z)]_{C_O=1};\qquad \hat{\Phi}_{-q}(z)=|\Phi_{-q}(z)]_{C_{-q}=1};$$

$$\hat{\Phi}_{-q}(z) = [\Phi_{-q}(z)]_{C_{-q}=\mathrm{i}},$$

the general solution (6.16) may be written in the form

$$\Phi^*(z) = c_O\hat{\Phi}_O(z) + \sum_{q=1}^{[|\frac{n}{2}|]} \hat{c}_{-q}\hat{\Phi}_{-q}(z) + \hat{\hat{c}}_{-q}\hat{\hat{\Phi}}_{-q}(z), \tag{6.17}$$

where $\hat{c}_{-q} = \mathrm{Re}\ c_{-q}, \hat{\hat{c}}_{-q} = \mathrm{Im} c_{-q}$.

Therefore the homogeneous boundary problem (6.1), (6.2) has n linearly independent solutions over the field of real numbers and the aggregate of all solutions is given by any of the formulas (6.16) or (6.17).

Now we examine the inhomogeneous problem. Without dwelling upon intermediate calculations, we shall just indicate that $w(z)$, defined by the equality

$$w(z) = \sum_{s=0}^{\infty} w_s(z), \tag{6.18}$$

gives a particular solution of our problem. Here $w_s(z)$ is the solution of equation (6.1) with the boundary condition of the particular kind:

$$\mathrm{Re}\ w_s = h_s \mathrm{e}^{\mathrm{i}s\varphi} + \bar{h}_s \mathrm{e}^{\mathrm{i}(n-s)\varphi}, \quad \text{on } \Gamma.$$

Clearly, $w_s(z)$ can be represented by the relation

$$\begin{aligned} w_s(z) =& r^{\frac{n}{2}-\nu} \sum_{k=O}^{\infty} A_{-nk+s} K_{p_{-nk+s}}\left(\frac{|\lambda|}{2\nu} r^{-2\nu}\right) \mathrm{e}^{-\mathrm{i}(nk+s)\varphi} \\ &+ \frac{\lambda}{|\lambda|} \bar{A}_{-nk+s} K_{p_{-nk+s}-1}\left(\frac{|\lambda|}{2\nu} r^{-2\nu}\right) \mathrm{e}^{\mathrm{i}[n(k+1)-s]\varphi}. \end{aligned}$$

Here for all $s = 0, 1, \ldots,$ we have

$$A_s = \frac{R^{\nu-\frac{n}{2}}}{K_{p_s}\left(\frac{|\lambda|}{2\nu} R^{-2\nu}\right)} h_s$$

and

$$A_{-nk+s} = \left(-\frac{\bar{\lambda}}{|\lambda|}\right)^k \prod_{m=1}^{k} \frac{K_{p_{nm-s}}\left(\frac{|\lambda|}{2\nu} R^{-2\nu}\right)}{K_{p_{nm+s}}\left(\frac{|\lambda|}{2\nu} R^{-2\nu}\right)} \cdot \frac{R^{\nu-\frac{n}{2}}}{K_{p_s}\left(\frac{|\lambda|}{2\nu} R^{-2\nu}\right)} h_s.$$

One establishes that

$$|w_s| < R^{\nu - \frac{n}{2}} \cdot \frac{2}{1-q} |h_s|,$$

where $0 < q < 1$. Hence and from (6.15) it follows that

$$|w(z)| \leq \sum_{s=0}^{\infty} |w_s(z)| \leq R^{\nu - \frac{n}{2}} \frac{2}{1-q} \sum_{s=0}^{\infty} |h_s|.$$

This guarantees the absolute and uniform convergence of series (6.15). Thus, Theorem 6.1, with odd n is entirely proved.

In a similar way the statement of the theorem for even n is verified.

In the case of $n \geq 0$ the situation is more complicated. The solution of the system of algebraic equations (6.8)–(6.10) is expressed by n arbitrary constants. But in future they will not remain arbitrary. Their values should be fixed in a unique way to provide the absolute and uniform convergence of (6.3), (6.4). But this is not sufficient. To prove the series convergence $h(\varphi) = h(Re^{i\varphi})$ should be subordinated to additional conditions. Without dwelling on the proof of the solvability of (6.1), (6.2) with $n = 0$ and $n > 0$ we formulate the final results.

Theorem 6.1 (continuation) *Let $n = 0$. If λ is a negative number, then the homogeneous problem (6.1), (6.2) has one non-trivial solution, and the inhomogeneous problem is solvable if and only if*

$$\int_0^{2\pi} h(Re^{i\varphi}) \, d\varphi = 0.$$

For other values of λ a solution of the boundary problem exists and is unique.

Let $n \geq 1$. Then the homogeneous boundary problem has only the trivial solution. A solution of the inhomogeneous problem exists if n solvability conditions are satisfied:

$$h_O + \mathrm{Re}\left\{ \sum_{\delta=1}^{\infty} \left(-\frac{\bar{\lambda}}{|\lambda|} \right)^{\delta} \frac{\prod_{m=1}^{\delta} K_{p_{n(1-m)}}\left(\frac{|\lambda|}{2\nu} R^{-2\nu} \right)}{\prod_{m=1}^{\delta+1} K_{p_{n(m-1)}}\left(\frac{|\lambda|}{2\nu} R^{-2\nu} \right)} h_{ns} \right\} = 0 \qquad (6.19)$$

$$K_{p_s}\left(\frac{|\lambda|}{2\nu} R^{-2\nu} \right) h_s + \frac{K_{p_{-s}}\left(\frac{|\lambda|}{2\nu} R^{-2\nu} \right)}{K_{p_s}\left(\frac{|\lambda|}{2\nu} R^{-2\nu} \right)} \sum_{\delta=1}^{\infty} \left(-\frac{\bar{\lambda}}{|\lambda|} \right)^{\delta}$$

$$\times \frac{\prod_{m=1}^{\delta} K_{p_{n(1-m)+s}}\left(\frac{|\lambda|}{2\nu}R^{-2\nu}\right)}{\prod_{m=1}^{\delta+1} K_{p_{n(m-1)-s}}\left(\frac{|\lambda|}{2\nu}R^{-2\nu}\right)}\bar{h}_{n\delta-s}$$

$$+\sum_{\delta=1}^{\infty}\left(-\frac{\bar{\lambda}}{|\lambda|}\right)^{\delta} \frac{\prod_{m=1}^{\delta} K_{p_{n(1-m)-s}}\left(\frac{|\lambda|}{2\nu}R^{-2\nu}\right)}{\prod_{m=1}^{\delta+1} K_{p_{n(m-1)+s}}\left(\frac{|\lambda|}{2\nu}R^{-2\nu}\right)} h_{n\delta+s} = 0 \tag{6.20}$$

$$(s = 1, 2, \ldots, n_O - 1).$$

For even n it is necessary to add the following solvability condition

$$\mathrm{Re}\left\{ \frac{h_{\frac{n}{2}}}{K_{\frac{1}{2}}\left(\frac{|\lambda|}{2\nu}R^{-2\nu}\right)} + \sum_{\delta=1}^{\infty}\left(-\frac{\bar{\lambda}}{|\lambda|}\right)^{\delta} \frac{\prod_{m=1}^{\delta} K_{p_{n(1-2m)/2}}\left(\frac{|\lambda|}{2\nu}R^{-2\nu}\right)}{\prod_{m=1}^{\delta+1} K_{p_{n(2m-1)/2}}\left(\frac{|\lambda|}{2\nu}R^{-2\nu}\right)} \right.$$

$$\left. \times h_{(2\delta+1)n/2} \right\} - \frac{\mathrm{Im}\lambda}{|\lambda| - \mathrm{Re}\lambda} \mathrm{Im}\left\{ \frac{h_{\frac{n}{2}}}{K_{\frac{1}{2}}\left(\frac{|\lambda|}{2\nu}R^{-2\nu}\right)} \right.$$

$$\left. +\sum_{\delta=1}^{\infty}\left(-\frac{\bar{\lambda}}{|\lambda|}\right)^{\delta} \frac{\prod_{m=1}^{\delta} K_{p_{n(1-2m)/2}}\left(\frac{|\lambda|}{2\nu}R^{-2\nu}\right)}{\prod_{m=1}^{\delta+1} K_{p_{n(2m-1)/2}}\left(\frac{|\lambda|}{2\nu}R^{-2\nu}\right)} .h_{(2\delta+1)n/2} \right\} = 0.$$

to the previous relations. In addition, if λ is a positive number, then the solvability condition is simplified:

$$\mathrm{Im}\left\{ \frac{h_{\frac{n}{2}}}{K_{\frac{1}{2}}\left(\frac{|\lambda|}{2\nu}R^{-2\nu}\right)} \right.$$

$$\left. +\sum_{\delta=1}^{\infty}\left(-\frac{\bar{\lambda}}{|\lambda|}\right)^{\delta} \frac{\prod_{m=1}^{\delta} K_{p_{n(1-2m)/2}}\left(\frac{|\lambda|}{2\nu}R^{-2\nu}\right)}{\prod_{m=1}^{\delta+1} K_{p_{n(2m-1)/2}}\left(\frac{|\lambda|}{2\nu}R^{-2\nu}\right)} .h_{(2\delta+1)n/2} \right\} = 0.$$

If λ is a negative number then it takes the form

$$\mathrm{Re}\left\{ \frac{h_{\frac{n}{2}}}{K_{\frac{1}{2}}\left(\frac{|\lambda|}{2\nu}R^{-2\nu}\right)} \right.$$

$$+\sum_{\delta=1}^{\infty}\left(-\frac{\bar{\lambda}}{|\lambda|}\right)^{\delta}\frac{\prod_{m=1}^{\delta} K_{p_{n(1-2m)/2}}\left(\frac{|\lambda|}{2\nu}R^{-2\nu}\right)}{\prod_{m=1}^{\delta+1} K_{p_{n(2m-1)/2}}\left(\frac{|\lambda|}{2\nu}R^{-2\nu}\right)}.h_{(2\delta+1)n/2}\Bigg\} = 0.$$

6.2 Riemann–Hilbert problem *It is necessary to find a solution $\Phi(z) \in C(\bar{G}) \cap C_{\bar{z}}(G-0)$ of equation (6.1) in the domain $G = \{z : |z| < R\}$, satisfying the boundary condition*

$$\mathrm{Re}(z^{-m}\Phi) = h(z) \quad \text{on } \Gamma, \tag{6.21}$$

where m is an integer, $\Gamma = \{z : z = Re^{i\varphi}\}$ is a circle and $h(z) = h(Re^{i\varphi})$ as in subsection 6.1 is expanded in an absolutely convergent Fourier series.

The analysis of this problem with $m = 0$ (the Dirichlet problem) is carried out in subsection 6.1.

We formulate the solvability results of boundary problem (6.1), (6.21) for different combinations of n and m. As before, the boundary problem will be called homogeneous if $h(z) \equiv 0$.

Theorem 6.2 *Let $n < 2m$. Then the homogeneous problem has $2m - n$ linearly independent solutions. The inhomogeneous problem is undoubtedly solvable.*

Let $n = 2m$. If λ is a negative number, then the homogeneous problem has one non-trivial solution, and to solve the inhomogeneous problem it is necessary and sufficient to have $h_O = 0$. For other values of λ a solution of the problem exists and is unique.

Let $n > 2m$. Then the homogeneous problem has only the trivial solution. A solution of the inhomogeneous problem exists if $n - 2m$ solvability conditions are fulfilled.

Proof Let us introduce the unknown function $\Phi^O(z) = z^{-m}\Phi(z)$. It will evidently satisfy the problem

$$\partial_{\bar{z}}\Phi^O - \frac{\lambda}{2r^{1+2\nu}}e^{i(n-2m+1)\varphi}\bar{\Phi}^O = 0, \quad z \in G, \tag{6.22}$$

$$\mathrm{Re}\ \Phi^O(z) = h(z), \quad z \in \Gamma. \tag{6.23}$$

Considering Theorem 5.3, any solution of equation (6.1) from the class $C(\bar{G}) \cap C_{\bar{z}}(G-0)$ has a zero of an arbitrarily high order at $z = 0$, so $\Phi^O(z)$ also belongs to the same class. But then the results of Theorem 6.1 can spread to problem (6.22), (6.23). This is what makes Theorem 6.2 correct. The solvability condition which were described in the theorem are easily obtained from (6.19), (6.20).

6.3 The linear conjugation problem Let $\Phi^+(z) \in C(G^+ + \Gamma) \cap C_{\bar{z}}(G^+ - 0)$ satisfy the equation

$$\partial_{\bar{z}}\Phi^+ - \frac{\lambda}{2r^{1+2\nu}}\mathrm{e}^{\mathrm{i}(n+1)\varphi}\bar{\Phi}^+ = 0, \quad z \in G^+, \tag{6.24}$$

where $G^+ = \{z : |z| < R\}$ is a disk and $\Gamma = \{z : z = R\mathrm{e}^{\mathrm{i}\varphi}\}$ is its boundary. We denote an analytic function in the domain $G^- = \{z : |z| > R\}$ by $\Phi(z)$ and assume that $\Phi(z)$ is bounded at $z = \infty$ and continuously extendable on Γ.

Problem *It is necessary to determine functions* $\Phi^+(z), z \in G^+, \Phi^-(z), z \in G^-$, *under the condition*

$$\Phi^+(z) = z^m\Phi^-(z) + g(z), \quad z \in \Gamma. \tag{6.25}$$

We assume that the function $g(z) = g(R\mathrm{e}^{\mathrm{i}\varphi})$ is expanded in an absolutely convergent Fourier series:

$$g(z) = g_O + \sum_{k=1}^{\infty} g_k\mathrm{e}^{\mathrm{i}k\varphi} + g_{-k}\mathrm{e}^{-\mathrm{i}k\varphi},$$

where

$$g_k = \frac{1}{2\pi}\int_0^{2\pi} g(R\mathrm{e}^{\mathrm{i}\varphi})\mathrm{e}^{-\mathrm{i}k\varphi}\mathrm{d}\varphi, \quad k = 0, \pm 1, \ldots .$$

Theorem 6.3 *Let* n *be an odd number. If* $m > n_O$ *then the pair of functions* $\Phi_O^\pm(z)$, *forming a solution of the homogeneous problem* ($g(z) \equiv O$), *contains* $2(m - n_O + 1)$ *arbitrary real constants and is given by the formulas:*

$$\begin{aligned}\Phi_O^+(z) =& r^{\frac{n}{2}-\nu}\sum_{k=n_o}^{m} B_k K_{p_k}\left(\frac{|\lambda|}{2\nu}r^{-2\nu}\right)\mathrm{e}^{\mathrm{i}k\varphi} \\ &+ \frac{\lambda}{|\lambda|}\bar{B}_k K_{p_k-1}\left(\frac{|\lambda|}{2\nu}r^{-2\nu}\right)\mathrm{e}^{\mathrm{i}(n-k)\varphi}, \\ \Phi_O^-(z) =& R^{\frac{n}{2}-\nu-m}\sum_{k=n_o}^{m} B_k K_{p_k}\left(\frac{|\lambda|}{2\nu}R^{-2\nu}\right)\left(\frac{R}{Z}\right)^{m-k} \\ &+ \frac{\lambda}{|\lambda|}\bar{B}_k K_{p_k-1}\left(\frac{|\lambda|}{2\nu}r^{-2\nu}\right)\left(\frac{R}{Z}\right)^{m+k-n}.\end{aligned}$$

The inhomogeneous problem is always solvable.

If $m < n_O$, then the homogeneous problem has only the zero solution and to solve the inhomogeneous problem it is necessary and sufficient that $g(z)$ should satisfy $2(n_O - m - 1)$ real solvability conditions:

$$|\lambda| K_{p_k - 1}\left(\frac{|\lambda|}{2\nu} R^{-2\nu}\right) g_k - \lambda K_{p_k}\left(\frac{|\lambda|}{2\nu} R^{-2\nu}\right) \bar{g}_{n-k} = 0 \tag{6.26}$$

$$(k = n_O, n_O + 1, \ldots n - m - 1).$$

Let n be an even number. For $m \geq n_O$ the pair of functions $\Phi_O^{\pm}(z)$, forming a solution of the homogeneous boundary problem, contains $2(m - n_O) + 1$ arbitrary real constants and is given by the formulas:

$$\begin{aligned}
\Phi_O^+(z) =& r^{\frac{n}{2} - \nu}\Bigg[B_{n_O} K_{\frac{1}{2}}\left(\frac{|\lambda|}{2\nu} r^{-2\nu}\right) e^{i n_o \varphi} \\
&+ \sum_{k=n_o-1}^{m} B_k K_{p_k}\left(\frac{|\lambda|}{2\nu} r^{-2\nu}\right) e^{ik\varphi} + \frac{\lambda}{|\lambda|} \bar{B}_k K_{p_k - 1}\left(\frac{|\lambda|}{2\nu} r^{-2\nu}\right) e^{i(n-k)\varphi}\Bigg], \\
\Phi_O^-(z) =& r^{\frac{n}{2} - \nu - m}\Bigg[B_{n_O} K_{\frac{1}{2}}\left(\frac{|\lambda|}{2\nu} r^{-2\nu}\right)\left(\frac{R}{z}\right)^{m - n_o} \\
&+ \sum_{k=n_o+1}^{m} B_k K_{p_k}\left(\frac{|\lambda|}{2\nu} R^{-2\nu}\right)\left(\frac{R}{z}\right)^{m-k} + \frac{\lambda}{|\lambda|} \bar{B}_k K_{p_k - 1}\left(\frac{|\lambda|}{2\nu} R^{-2\nu}\right)\left(\frac{R}{z}\right)^{m+k-n}\Bigg],
\end{aligned}$$

and $|\lambda| B_{n_o} = \lambda \bar{B}_{n_o}$, and the inhomogeneous problem (6.24), (6.25) always has a solution.

For $m < n_O$ the homogeneous problem has only the zero solution and to solve the inhomogeneous problem it is necessary and sufficient that $g(z)$ should satisfy $2(n_O - m) - 1$ solvability conditions of type (6.26).

The proof of this theorem is as follows. A solution $\Phi^+(z)$ of (6.24) in the series form (6.3) or (6.4) and an analytic function $\Phi^-(z)$ represented by the power series with respect to $z \in G^-$, are substituted into the conjugation condition (6.25). Then we determine the arbitrary constants, form solutions, prove their convergence, etc.

7 Properties of the solutions of the general equation at a singular point

Now that certain information about the solutions of the model equation has been obtained in §§5,6 we shall turn again to the general case. In the present section the object of our investigation is the equation

$$\partial_{\bar{z}} w - \frac{b(z)}{2r^{1+2\nu}} e^{i(n+1)\varphi} \bar{w} = F(z), \quad z \in G, \tag{7.1}$$

with the same suppositions with respect to $b(z)$ that were mentioned in the introduction to this chapter.

Theorem 7.1 *Let $w(z)$ be a solution of equation (7.1) from the class $C(\bar{G}) \cap G_{\bar{z}}(G - 0)$. Then $w(z)$ has a zero of an arbitrary high order at the point $z = 0$.*

Theorem 7.2 *Let $b(z) \in C^m(\bar{G}), m \geq 1$. Then a solution $w(z)$ from the class $C(\bar{G}) \cap C_{\bar{z}}(G - 0)$ belongs to the class $C^m(\bar{G})$.*

To prove Theorem 7.1 one should turn to formula (4.2), establishing the connection between the sets of solutions $\{w(z)\}$ and $\{\Phi(z)\}$ of the general and model equations, add the information of $\Phi(z)$ having a zero of an arbitrarily high order at $z = 0$ (Theorem 5.3) to that and then repeat the scheme of proving Theorem 1.1 of ch. 3.

We also do not dwell on verifying Theorem 7.2 since it is similar to verifying Theorem 5.3 of this chapter.

8 The Riemann–Hilbert problem for solutions of the general equation

Let G be a simply connected domain of a complex plane z bounded by the Ljapunov contour Γ and containing the origin in its interior.

Problem: *It is necessary to find a function $w(z) \in C(\bar{G}) \cap C_{\bar{z}}(G - 0)$ subordinated to the demands:*

$$\partial_{\bar{z}} w - \frac{b(z)}{2r^{1+2\nu}} \mathrm{e}^{\mathrm{i}(n+1)\varphi} \bar{w} = F(z), \quad z \in G, \tag{8.1}$$

$$\mathrm{Re}\, \{\overline{g(z))} w(z)\} = h(z), \quad z \in \Gamma \tag{8.2}$$

where the functions $g(z)$ and *$h(z)$ are Holder continuous on Γ. In respect of $b(z)$* and *$F(z)$ one assumes the conditions formulated in the introduction and §4 of this chapter.*

As in §3, ch. 3, the problem (8.1), (8.2) can be essentially simplified. By a conformal map $\zeta = f(z)(f(0) = 0$ and $\arg f'(0) = 0)$ the domain G is transformed into the disk $|\zeta| < R$. Not only the form of the initial equation (8.1), but also the value of $b(z)$ at the origin, remains in the new variable. Further by means of a real regularizing multiplier and a new unknown function the boundary condition (8.2) can be reduced to the form

$$\mathrm{Re}\, [z^{-m} w] = h(z), \quad |z| = R, \tag{8.3}$$

where m is the index of $g(z)$. Again the equation in terms of the new unknown function will have the form (8.1). For this reason we can assume that $G = \{z : |z| < R\}$ is a circular domain, and we can examine the boundary problem (8.3) instead of (8.2).

The method of analysing problem (8.1), (8.3) is mainly similar to that given in §§4–6, ch. 3. It means that the two–dimensional integral equation (4.2), connecting the sets of solutions $w(z)$ and $\Phi(z)$ of the general and model equations, constitutes the basis for further investigation. The integral equation is modified so that the problem which interests us can be reduced to a Riemann–Hilbert boundary problem for solutions of the model equation.

The specificity of object (8.1) is shown in the fact that we should examine three cases ($n = 2m, n < 2m$ and $n > 2m$) and use the modified Bessel functions for constructing the kernels of additional integral operators. Expressions of these kernels prove to be very bulky, so we do not write them here. For more detailed information see [41, 42]; or the expressions can be obtained independently, applying the argument of §8 ch. 3.

8.1 The case $n = 2m$ Let us deduce a necessary condition for solvability with $b(0) < 0$. To this end we write (8.1) in the form

$$\partial_{\bar{z}} w - \frac{\lambda}{2r^{1+2\nu}} \mathrm{e}^{\mathrm{i}(n+1)\varphi} \bar{w} = f(z), \tag{8.4}$$

where $\lambda = b(0)$ and

$$f(z, w) = F(z) + \frac{b(z) - b(0)}{2r^{1+2\nu}} \mathrm{e}^{\mathrm{i}(n+1)\varphi} \overline{w(z)}.$$

If one assumes that f is a known function, then the homogeneous boundary problem, conjugated to (8.1), (8.3), is the following:

$$\partial_{\bar{z}} U - \frac{\bar{\lambda}}{2r^{1+2\nu}} \mathrm{e}^{-\mathrm{i}(n+1)\varphi} \bar{U} = 0, \quad |z| < R, \tag{8.5}$$

$$\mathrm{Re}[\mathrm{i}\mathrm{e}^{\mathrm{i}(m+1)\varphi} U] = 0, \quad |z| = R. \tag{8.6}$$

As was shown in §6, this problem has a non–zero solution of the class $C(\bar{G}) \cap G_{\bar{z}}(G - 0)$:

$$U(z) = r^{-\nu-m-1} K_{\frac{1}{2}}\left(\frac{|\lambda|}{2\nu} r^{-2\nu}\right) \mathrm{e}^{-\mathrm{i}(m+1)\varphi},$$

where $K_{1/2}(t)$ is a MacDonald function. Let us apply Green's identity to the solutions of (8.4) and (8.5):

$$\mathrm{Re}\left[\frac{1}{2\mathrm{i}}\int_{\Gamma} wU\,\mathrm{d}z\right] = \mathrm{Re}\iint\limits_{G} Uf\,\mathrm{d}x\,\mathrm{d}y, \quad z = x + \mathrm{i}y.$$

If we use the expressions for U and f then we get

$$T_G\left(\frac{b(z)-b(0)}{2r^{1+2\nu}}\mathrm{e}^{2\mathrm{i}m\varphi}\overline{w(z)}\right) = h_O - T_G\left(2\mathrm{e}^{-\mathrm{i}\varphi}F(z)\right). \tag{8.7}$$

Here

$$T_G f = \frac{R^{\nu}}{2\pi K_{\frac{1}{2}}\left(\frac{|\lambda|}{2\nu}R^{-2\nu}\right)}\mathrm{Re}\iint\limits_{G} r^{-\nu-m-1}K_{\frac{1}{2}}\left(\frac{|\lambda|}{2\nu}r^{-2\nu}\right)f(z)\mathrm{e}^{-\mathrm{i}m\varphi}\,\mathrm{d}x\,\mathrm{d}y$$

is a linear bounded functional in $L_p^*, p > 2$, and

$$h_O = \frac{1}{2\pi}\int_O^{2\pi} h(R\mathrm{e}^{\mathrm{i}\theta})\,\mathrm{d}\theta.$$

Thus, if $w(z)$ is a solution of problem (8.1), (8.3) with $n = 2m$ and $\lambda = b(0) < 0$, then it must satisfy condition (8.7).

Now we introduce the auxiliary integral operator

$$S_G^O f = -\frac{1}{\pi}\iint\limits_{G}\left[\frac{\Omega_1^O(z,\zeta,m)}{\zeta}f(z) + \frac{\Omega_2^O(z,\zeta,m)}{\bar{\zeta}}\overline{f(z)}\right]\mathrm{d}\xi\,\mathrm{d}\eta,$$

which proves to be completely continuous from $L_p^*(\bar{G}), p > 2$, in $\bar{C}(G)$ and

$$\|S_G^O f\|_C \le N^O R^{\mathcal{X}}\|f\|_{L_p^*},$$

where N^O and $\mathcal{X}$ are fully definite positive constants, $\|f\|_{L_p^*} = \|f_O\|_{L_p}, f(z) = r^{4\nu}f_O(z)$. The function $S_G^O f$ is a continuous solution of the model equation

$$\partial_{\bar{z}}\Phi - \frac{\lambda}{2r^{1+2\nu}}\mathrm{e}^{\mathrm{i}(n+1)\varphi}\bar{\Phi} = 0, \quad z \in G. \tag{8.8}$$

Together with $S_G f$, given by (1.11), $S_G^O f$ satisfies the condition

$$\mathrm{Re}\left[z^{-m}(S_G f + S_G^O f)\right]_{|z|=R} = \begin{cases} T_G(2\mathrm{e}^{\mathrm{i}\psi}f(\zeta)), & \lambda < 0, \\ 0 & \text{for other } \lambda. \end{cases} \tag{8.9}$$

One more auxiliary operator

$$P_G^O \bar{w} = S_G^O\left(\frac{b(\zeta)-b(0)}{2\rho^{1+2\nu}} e^{i(2m+1)\psi}\overline{w(\zeta)}\right).$$

is also completely continuous from $C(\bar{G})$ to itself, in addition

$$\|P_G^O\| \leq N^O R^{\varkappa}\left\|\frac{b(z)-b(0)}{2\rho^{6\nu+1}}\right\|_{L_p}, \quad p>2. \tag{8.10}$$

The function $P_G^O \bar{w}$ is a continuous solution of equation (8.8), jointly with $P_G w$, described in §4; it satisfies the equality

$$\operatorname{Re}\left\{z^{-m}(P_G\bar{w}+P_G^O w)\right\}_{|z|=R} = \begin{cases} T_G\left(\frac{b(\zeta)-b(0)}{2\rho^{1+2\nu}}\overline{w(\zeta)}\right) & \text{for } \lambda>0, \\ 0 & \text{for other } \lambda. \end{cases} \tag{8.11}$$

Now for analysing the boundary problem we turn to the general equation (4.2):

$$w(z) = \Phi(z) + S_G F + P_G \bar{w}. \tag{8.12}$$

Representing $\Phi(z)$ of equation (8.8) in the form

$$\Phi(z) = \Phi_O(z) + S_G^O F + P_G^O \bar{w},$$

where $\Phi_O(z)$ is a new unknown solution of (8.8) from $C(\bar{G}) \cap C_{\bar{z}}(G-0)$, and introducing it into the right side of (8.12), we get

$$w(z) - \hat{P}_G \bar{w} = \Phi_O(z) + \hat{S}_G F \tag{8.13}$$

where $\hat{P}_G = P_G + P_G^O$ and $\hat{S}_G = S_G + S_G^O$.

If we multiply both sides of (8.13) by z^{-m}, and assume $|z| = R$ and take the real part, then thanks to boundary condition (8.3), we get

$$h(z) - \operatorname{Re}[z^{-m}\hat{P}_G\bar{w}] = \operatorname{Re}[z^{-m}\Phi_O(z)] + \operatorname{Re}[z^{-m}\hat{S}_G F] \quad \text{on } \Gamma. \tag{8.14}$$

Let $\lambda = b(0)$ be a non-negative number. Then taking relations (8.9), (8.11) into account we obtain from (8.14)

$$\operatorname{Re}[z^{-m}\Phi_O(z)] = h(z), \quad z \in \Gamma. \tag{8.15}$$

Hence one can see that $\Phi_O(z)$ is a continuous solution of boundary problem (8.8), (8.15), which according to Theorem 6.2 has a unique solution. Substituting this solution into (8.13), we arrive at an integral equation of Fredholm type, that is equivalent to boundary prolem (8.1) (8.3). If the condition $\|\hat{P}_G\|_C < 1$ is imposed then the problem has a unique solution.

Now let $\lambda = b(0) < 0$. Employing (8.9), (8.11) in (8.14) we get

$$\operatorname{Re}[z^{-m}\Phi_O(z)] = h(z) - T_G\left(\frac{b(\zeta)-b(0)}{\rho^{1+2\nu}}\mathrm{e}^{2\mathrm{i}m\psi}\overline{w(\zeta)}\right) - T_G\left(2\mathrm{e}^{-\mathrm{i}\psi}F(z)\right), z \in \Gamma.$$

According to (8.7) this equality takes the form

$$\operatorname{Re}[z^{-m}\Phi_O(z)] = h(z) - h_O \quad \text{on } \Gamma. \tag{8.16}$$

From Theorem 6.2 ($n = 2m$), the solution of boundary problem (8.8), (8.16) is expressed by the formula

$$\Phi_O(z) = \mathrm{i}Cr^{m-\nu}K_{\frac{1}{2}}\left(\frac{|\lambda|}{2\nu}r^{-2\nu}\right)\mathrm{e}^{\mathrm{i}m\varphi} + H(z),$$

where the first summand is a solution of the homogeneous boundary problem (8.8), (8.16), and the second one

$$H(z) = \left(\frac{r}{R}\right)^{m-\nu}\sum_{k=1}^{\infty}\frac{K_{p_k}\left(\frac{|\lambda|}{2\nu}r^{-2\nu}\right)h_k\mathrm{e}^{\mathrm{i}k\varphi} + \frac{\lambda}{|\lambda|}K_{p_k-1}\left(\frac{|\lambda|}{2\nu}r^{-2\nu}\right)\bar{h}_k\mathrm{e}^{-\mathrm{i}k\varphi}}{K_{p_k}\left(\frac{|\lambda|}{2\nu}R^{-2\nu}\right) + \frac{\lambda}{|\lambda|}K_{p_k-1}\left(\frac{|\lambda|}{2\nu}R^{-2\nu}\right)}$$

is a solution of the inhomogeneous problem. Therefore (8.13) take the form

$$w(z) - \hat{P}_G\bar{w} = \mathrm{i}Cr^{m-\nu}K_{\frac{1}{2}}\left(\frac{|\lambda|}{2\nu}r^{-2\nu}\right)\mathrm{e}^{\mathrm{i}m\varphi} + H(z) + \hat{S}_GF. \tag{8.17}$$

Thus, boundary problem (8.1), (8.3) with $n = 2m$ and $\lambda = b(0) < 0$ is equivalent to integral equation (8.17) with additional condition (8.7).

Let $\|\hat{P}_G\|_C < 1$. Because of the linearity of equation (8.17) its solutions are represented in the form $w = w_O + Cw_1$, where w_O and w_1 are, respectively, solutions of the equations:

$$w_O - \hat{P}_G\bar{w}_O = H(z) + \hat{S}_GF,$$
$$w_1 - \hat{P}_G\bar{w}_1 = \mathrm{i}r^{m-\nu}K_{\frac{1}{2}}\left(\frac{|\lambda|}{2\nu}r^{-2\nu}\right)\mathrm{e}^{\mathrm{i}m\varphi}.$$

Then (8.7) can be rewritten as follows:

$$CT_G\left(\frac{b(z|) - b(0)}{r^{1+2\nu}} e^{2im\varphi}\overline{w_1(z)}\right) = h_O - T_G\left(2e^{-i\varphi}F(z) + \frac{b(z) - b(0)}{r^{1+2\nu}} e^{2im\varphi}\overline{w_O(z)}\right).$$

The realization of this equality in choosing a value for C depends on

$$\tau = T_G\left(\frac{b(z) - b(0)}{r^{1+2\nu}} e^{2im\varphi}\overline{w_1(z)}\right).$$

If $\tau \neq 0$, then C is defined uniquely and problem (8.1), (8.3) with $n = 2m$ has a unique solution. When $\tau = 0$, then the homogeneous problem ($F = h = 0$) will have a non–trivial solution while the inhomogeneous problem is solvable with fulfillment of the condition

$$T_G\left(2e^{-i\varphi}F(z) + \frac{b(z) - b(0)}{r^{1+2\nu}} e^{2im\varphi}\overline{w_0(z)}\right) = h_O. \tag{8.18}$$

Thus, we have established

Theorem 8.1 ($n = 2m$) *Let $\|\hat{P}_G\|_C < 1$. Then for $b(0) > 0$ or $\mathrm{Im}b(0) \neq 0$, and for $b(0) < 0$ with $\tau \neq 0$ as well, the solution of boundary problem (8.1), (8.3) exists and is unique.*

For $b(0) < 0$ with $\tau = 0$ the homogeneous problem ($F = h = 0$) has a non–trivial solution while for the existence of a solution of the inhomogeneous problem it is necessary and sufficient to satisfy condition (8.18).

If the inequality $\|\hat{P}_G\|_C < 1$ is not fulfilled then a more general statement will be valid.

Theorem 8.2 ($n = 2m$) *The index $\mathcal{X}$ of problem (8.1), (8.3) is equal to zero.*

Here $\mathcal{X} = l - l'$, where l is the number of linearly independent solutions of the homogeneous boundary problem and l' is the number of real solvaility conditions of the inhomogeneous problem.

8.2 The case $n < 2m$ As in the previous subsection, the important point of studying the boundary problem with $n < 2m$ is the construction of an auxilliary integral operator, by which the analysis of this problem is reduced to a similar problem for model equation (8.8).

Let us introduce the operators

$$S_G^+ f = -\frac{1}{\pi}\iint\limits_G \left[\frac{\Omega_1^+(z,\zeta,m)}{\zeta} f(\zeta) + \frac{\Omega_2^+(z,\zeta,m)}{\bar{\zeta}}\overline{f(\zeta)}\right] d\xi\, d\eta,$$

$$P_G^+\bar{w} = S_G^+\left(\frac{b(\zeta)-b(0)}{2\rho^{1+2\nu}}e^{i(n+1)\psi}\overline{w(\zeta)}\right).$$

with the same properties which have been enumerated in subsection 8.1 for the operators S_G^O and P_G^O and instead of (8.9), (8.11) the following equalities hold:

$$\mathrm{Re}[z^{-m}(S_G f + S_G^+ f)] = 0,$$

$$\mathrm{Re}[z^{-m}(P_G\bar{w} + P_G^+\bar{w})] = 0, \quad \text{on } \Gamma.$$

We should note that the explicit forms of Ω_1^+, Ω_2^+ are given in [41].

We represent $\Phi(z)$ in the form (8.12)

$$\Phi(z) = \Phi_O(z) + S_G^+ F + P_G^+\bar{w},$$

where $\Phi_O(z)$ is a new sought solution of model equation (8.8). Then we get

$$w(z) - P_G^*\bar{w} = \Phi_O(z) + S_G^* F, \tag{8.19}$$

$P_G^* = P_G + P_G^+$ and $S_G^* = S_G + S_G^+$, here P_G^+ is a linear completely continuous operator mapping the space $C(G)$ into itself and subordinated to an inequality of type (8.10).

Introducing $w(z)$ from (8.19) into (8.3) and taking the properties of $S_G^+ f, P_G^+\bar{w}$ on Γ into consideration we arrive at the condition

$$\mathrm{Re}[z^{-m}\Phi_O(z)] = h(z) \quad \text{on } \Gamma. \tag{8.20}$$

Consequently, any continuous solution of integral equation (8.19) will automatically satisfy (8.3), if $\Phi_O(z)$ is a solution of boundary problem (8.8), (8.20). The general solution of this problem is written down in §6, and should be substituted for $\Phi_O(z)$ in (8.19).

Theorem 8.1 ($n < 2m$) *Boundary problem* (8.1), (8.3) *in the class of continuous functions is equivalent to integral equation* (8.19).

If $\|P_G^*\|_C < 1$. *Then the homogeneous problem has* $2m - n$ *linearly independent solutions while the inhomogeneous problem is solvable with any* $h(z)$ *and* $F(z)$.

In the more general case we have

Theorem 8.2 ($n < 2m$) *The index* $\mathcal{X}$ *of problem (8.1), (8.3) is equal to* $2m - n$.

8.3 The case $n > 2m$ The necessary solvability conditions of the problem are derived as in the case $n = 2m$ and have the following form:

$$T_q\left(\frac{b(\zeta)-b(0)}{r^{1+2\nu}}e^{i(n+1)\varphi}\overline{w(\zeta)}\right)+T_q(F(z)) = B_q, q = 0,1,\dots,\left|\left[\frac{2m-n+1}{2}\right]\right| \quad (8.21)$$

where

$$B_O = \frac{1}{2\pi}\int_O^{2\pi} e^{i(m+1)\varphi}U_O(Re^{i\varphi})h(\varphi)\,d\varphi,$$

$$B_q = \frac{1}{2\pi}\int_O^{2\pi} e^{i(m+1)\varphi}\left[U_q(Re^{i\varphi}) + iU_q^*(Re^{i\varphi})\right]h(\varphi)\,d\varphi,$$

$$T_O(g) = \frac{R^{-m-1}}{\pi}\iint_G \mathrm{Re}[U_O(z)g(z)]\,dx\,dy,$$

$$T_q(g) = \frac{R^{-m-1}}{\pi}\iint_G \left\{\mathrm{Re}[U_q(z)g(z)] + i\mathrm{Re}[U_q^*(z)g(z)]\right\}dxdy$$

$$q = 0,1,\dots,\left|[m - \frac{n-1}{2}]\right|, \quad \text{if } n \text{ is odd},$$

$$q = 0,1,\dots,\left|[m - \frac{n-1}{2}]\right| - 1, \quad \text{if } n \text{ is even},$$

and

$$B_{\frac{n}{2}-m} = \frac{1}{2\pi}\int_O^{2\pi} e^{i(m+1)\varphi}U_{\frac{n}{2}-m}(Re^{i\varphi})h(\varphi)\,d\varphi,$$

$$T_{\frac{n}{2}-m}(g) = \frac{R^{-m-1}}{\pi}\iint_G \mathrm{Re}[U_{\frac{n}{2}-m}(z)g(z)]\,dx\,dy,$$

where $U_O, U_{\frac{n}{2}-m}, U_q$ and U_q^* form a complete system of linearly independent solutions of problem (8.5), (8.6), the explicit form of which we do not write out here.

Thus, if $w(z)$ is a continuous solution of problem (8.1), (8.3) with $n > 2m$, then it must satisfy the condition (8.21).

Let us introduce the operators

$$S_G^- f = -\frac{1}{\pi}\iint_G \left[\frac{\Omega_1^-(z,\zeta,m)}{\zeta}f(\zeta) + \frac{\Omega_2^-(z,\zeta,m)}{\bar\zeta}\overline{f(\zeta)}\right]d\xi\,d\eta,$$

$$P_G^- \bar w = C_G^-\left(\frac{b(\zeta)-b(0)}{2\rho^{1+2\nu}}e^{2i(n+1)\psi}\overline{w(\zeta)}\right).$$

They possess the same properties as S_G^O, P_G^O from subsection 8.1.

In investigating boundary problem (8.1), (8.3) with $n > 2m$ we represent the function $\Phi(z)$ in the form

$$\Phi(z) = \Phi_O(z) + S_G^- F + P_G^- \bar{w},$$

where $\Phi_O(z)$ is a new unknown solution of model equation (8.8). Substituting $\Phi(z)$ in (8.12), we get

$$w(z) - P_G^{**}\bar{w} = \Phi_O(z) + S_G^{**}F \tag{8.22}$$

in which $P_G^{**} = P_G + P_G^-$ and $S_G^{**} = S_G + S_G^-$. The operator P_G^{**} completely continuously maps the space $C(\bar{G})$ into itself and is subjected to an inequality of type (8.10).

Introducing $w(z)$ from (8.22) into boundary condition (8.3) and taking into account

$$\left[z^{-m}(\Omega_1 + \Omega_1^-) + \overline{z^{-m}(\Omega_2 + \Omega_2^-)}\right] = \zeta R^{-m-1}[U_O(\zeta)\cos n\varphi$$

$$+\frac{1}{2}\sum_{k=1}^{|[m-\frac{n-1}{2}]|} U_k(\zeta)(e^{ik\varphi} + e^{i(n-k)\varphi}) - U_k^*(\zeta)(e^{ik\varphi} - e^{i(n-k)\varphi})\Big], |z| = R,$$

for odd n and

$$[z^{-m}(\Omega_1 + \Omega_1^-) + \overline{z^{-m}(\Omega_2 + \Omega_2^-)}] = \zeta R^{-m-1}[U_O(\zeta)\cos n\varphi + U_{\frac{n}{2}-m}(\zeta)\cos(m - \frac{n}{2})\varphi$$

$$+\frac{1}{2}\sum_{k=1}^{|[m-\frac{n-1}{2}]|-1} U_k(\zeta)(e^{ik\varphi} + e^{i(n-k)\varphi}) - U_k^*(\zeta)(e^{ik\varphi} - e^{i(n-k)\varphi})\Big], \quad |z| = R,$$

for even n, we come to the relation

$$\mathrm{Re}(z^{-m}\Phi_O) - \left\{T_O(F(z) + \frac{b(z) - b(0)}{2r^{1+2\nu}}e^{i(n+1)\varphi}\overline{w(z)}\right\}\cos n\varphi$$

$$-\sum_{k=1}^{|[m-\frac{n-1}{2}]|} \mathrm{Re}\left\{T_k(F(z) + \frac{b(z) - b(0)}{2r^{1+2\nu}}e^{i(n+1)\varphi}\overline{w(z)})\frac{e^{ik\varphi} + e^{i(n-k)\varphi}}{2})\right\}$$

$$-\sum_{k=1}^{|[m-\frac{n-1}{2}]|} \mathrm{Im}\left\{T_k(^*)\frac{e^{ik\varphi} + e^{i(n-k)\varphi}}{2}\right\} = h(z), \quad |z| = R,$$

if n is odd, and

$$\mathrm{Re}(z^{-m}\Phi_O) - \left\{T_O(F(z) + \frac{b(z)-b(0)}{2r^{1+2\nu}}\mathrm{e}^{\mathrm{i}(n+1)\varphi}\overline{w(z)})\right\}\cos n\varphi$$

$$-\left\{T_{\frac{n}{2}-m}(F(z) + \frac{b(z)-b(0)}{2r^{1+2\nu}}\mathrm{e}^{\mathrm{i}(n+1)\varphi}\overline{w(z)})\right\}\cos(\frac{n}{2}-m)\varphi$$

$$-\sum_{k=1}^{\|[m-\frac{n-1}{2}]\|-1}\left[\mathrm{Re}\left\{T_k(*)\frac{\mathrm{e}^{\mathrm{i}k\varphi}+\mathrm{e}^{\mathrm{i}(n-k)\varphi}}{2}\right\}\right.$$

$$\left.+\mathrm{Im}\left\{T_k(*)\frac{\mathrm{e}^{\mathrm{i}k\varphi}-\mathrm{e}^{\mathrm{i}(n-k)\varphi}}{2}\right\}\right] = h(z), \quad |z| = R,$$

if n is even. Thanks to (8.21) these relations lead to the boundary condition

$$\mathrm{Re}(z^{-m}\Phi_O) = h^*(z), \quad |z| = R, \tag{8.23}$$

where

$$h^*(z) = h(z) + B_O\cos n\varphi$$

$$+\sum_{k=1}^{\|[m-\frac{n-1}{2}]\|}\mathrm{Re}\left(B_k\frac{\mathrm{e}^{\mathrm{i}k\varphi}+\mathrm{e}^{\mathrm{i}(n-k)\varphi}}{2}\right) - \mathrm{Im}\left(B_k\frac{\mathrm{e}^{\mathrm{i}k\varphi}-\mathrm{e}^{\mathrm{i}(n-k)\varphi}}{2}\right)$$

for odd n, and

$$h^*(z) = h(z) + B_O\cos n\varphi + B_{\frac{n}{2}-m}\cos(\frac{n}{2}-m)\varphi$$

$$+\sum_{k=1}^{\|[m-\frac{n-1}{2}]\|}\mathrm{Re}\left(B_k\frac{\mathrm{e}^{\mathrm{i}k\varphi}+\mathrm{e}^{\mathrm{i}(n-k)\varphi}}{2}\right) - \mathrm{Im}\left(B_k\frac{\mathrm{e}^{\mathrm{i}k\varphi}-\mathrm{e}^{\mathrm{i}(n-k)\varphi}}{2}\right)$$

for even n.

The function $h^*(z)$ automatically satisfies the solvability conditions which are mentioned in Theorem 6.2 ($n > 2m$). So boundary problem (8.8), (8.23) has a unique solution. Denoting it by $H(z)$ and substituting for $\Phi_O(z)$ into (8.22), we get

$$w(z) - P_G^{**}w = H(z) + S_G^{**}F. \tag{8.24}$$

Theorem 8.1 ($n > 2m$) *The boundary problem is equivalent to integral equation (8.24) and additional conditions (8.21).*

For $\|P_G^{**}\|_C < 1$ the homogeneous problem (8.1), (8.3) ($F = h = 0$) has only the zero solution. For the existence of solutions of the inhomogeneous problem it *is necessary and sufficient that $F(z)$ and $h(z)$ should satisfy $n-2m$ real solvability conditions.*

Theorem 8.2 $(n>2m)$ *The index $\mathcal{X}$ of the boundary problem is equal to $2m-n$.*

9 The conjugation problem

Let $G^+=\{z:|z|<R\}, \Gamma=\{z:z=\mathrm{Re}^{\mathrm{i}\varphi}\}, G^-=\{z:|z|>R\}$.

Problem *It is necessary to find a solution $w^+(z)\in C(\overline{G^+})\cap C_{\bar z}(G^+-0)$ of the equation*

$$\partial_{\bar z}w^+ - \frac{b(z)}{2r^{1+2\nu}}\mathrm{e}^{\mathrm{i}(n+1)\varphi}\overline{w(z)}=F(z), \quad z\in G^+, \tag{9.1}$$

and an analaytic function $\Phi^-(z), z\in G^-$, continuous in $G^-+\Gamma$ and bounded at $z=\infty$, under the condition

$$w^+(z)=z^m\Phi^-(z)+g(z), \quad z\in\Gamma, \tag{9.2}$$

where $g(z)$ is a given function, Holder continuous on Γ. Concerning $b(z)$ and *$F(z)$ the same suppositions as before are accepted.*

As in the previous section the main point in the solution of this problem is the construction of an auxiliary integral operator and on the basis of this the transition to an integral equation equivalent to the boundary problem.

9.1 On the set $|z|,|\zeta|<R$ we introduce the functions Ω_1^+ and Ω_2^+ :

$$\Omega_1^+ = \frac{|\lambda|}{2\nu r^{2\nu}}\left(\frac{r}{\rho}\right)^{\nu+\frac{n}{2}}\sum_{k=n_O}^{\infty}\frac{I_{p_k}\left(\frac{|\lambda|}{2\nu}R^{-2\nu}\right)}{K_{p_k}\left(\frac{|\lambda|}{2\nu}R^{-2\nu}\right)}\left[K_{p_k}\left(\frac{|\lambda|}{2\nu}r^{-2\nu}\right)\right.$$

$$\left.\times K_{p_k-1}\left(\frac{|\lambda|}{2\nu}\rho^{-2\nu}\right)\mathrm{e}^{\mathrm{i}k(\varphi-\psi)} - K_{p_k-1}\left(\frac{|\lambda|}{2\nu}r^{-2\nu}\right)K_{p_k}\left(\frac{|\lambda|}{2\nu}\rho^{-2\nu}\right)\mathrm{e}^{\mathrm{i}(n-k)(\varphi-\psi)}\right]$$

$$\Omega_2^+ = \frac{|\lambda|\mathrm{e}^{\mathrm{i}n\psi}}{2\nu r^{2\nu}}\left(\frac{r}{\rho}\right)^{\nu+\frac{n}{2}}\sum_{k=n_O}^{\infty}\frac{I_{p_k}\left(\frac{|\lambda|}{2\nu}R^{-2\nu}\right)}{K_{p_k}\left(\frac{|\lambda|}{2\nu}R^{-2\nu}\right)}\left[K_{p_k}\left(\frac{|\lambda|}{2\nu}r^{-2\nu}\right)\right.$$

$$\left.\times K_{p_k-1}\left(\frac{|\lambda|}{2\nu}\rho^{-2\nu}\right)\mathrm{e}^{\mathrm{i}k(\varphi-\psi)} - K_{p_k-1}\left(\frac{|\lambda|}{2\nu}r^{-2\nu}\right)K_{p_k}\left(\frac{|\lambda|}{2\nu}\rho^{-2\nu}\right)\mathrm{e}^{\mathrm{i}(n-k)(\varphi-\psi)}\right).$$

We recall that $\lambda=b(0), n_O$ is the nearest integer to the left of $(n+1)/2, p_k=(k+\nu-\frac{n}{2})/2\nu$ and $I_p(t), K_p(t)$ are modified Bessel functions.

Let us consider the operator

$$S_G^+ f = -\frac{1}{\pi}\iint\limits_{G+}\left[\frac{\Omega_1^+(z,\zeta)}{\zeta}f(\zeta) + \frac{\Omega_2^{+-}(z,\zeta)}{\bar{\zeta}}\overline{f(\zeta)}\right] d\xi\, d\eta.$$

It is not valid to verify that S_G^+ is completely continuous from $L_q^*(G^+ + \Gamma), q > 2$, in $C(G^+ + \Gamma)$, and $S_G^+ f \in C(G^+ + \Gamma) \cap C_{\bar{z}}(G^+ - 0)$ and is a solution of model equation (8.8).

For $|z| > R$ and $|\zeta| < R$ we introduce two more pairs of functions $\overset{p}{\Omega}{}_1^-$ and $\overset{p}{\Omega}{}_2^-(p = 1, 2)$:

$$\overset{1}{\Omega}{}_1^- = -\left(\frac{R}{\rho}\right)^{\nu+\frac{n}{2}} \sum_{k=n_O}^{\infty} \left(\frac{K_{p_k}\left(\frac{|\lambda|}{2\nu}\rho^{-2\nu}\right)}{K_{p_k}\left(\frac{|\lambda|}{2\nu}R^{-2\nu}\right)}\left(\frac{R}{z}\right)^{k-n} e^{i(k-n)\psi},\right.$$

$$\overset{1}{\Omega}{}_2^- = \frac{\lambda}{|\lambda|}\left(\frac{R}{\rho}\right)^{\nu+\frac{n}{2}} \sum_{k=n_O}^{\infty} \frac{K_{p_k-1}\left(\frac{|\lambda|}{2\nu}\rho^{-2\nu}\right)}{K_{p_k}\left(\frac{|\lambda|}{2\nu}R^{-2\nu}\right)}\left(\frac{R}{z}\right)^{k-n} e^{ik\psi},$$

if n is odd, but if n is even then the summands with index $k = n_O$ should be divided by 2:

$$\overset{2}{\Omega}{}_1^- = -\left(\frac{R}{\rho}\right)^{\nu+\frac{n}{2}} \sum_{k=n-m}^{\infty} \frac{K_{p_k}\left(\frac{|\lambda|}{2\nu}\rho^{-2\nu}\right)}{K_{p_k}\left(\frac{|\lambda|}{2\nu}R^{-2\nu}\right)}\left(\frac{R}{z}\right)^{k-n} e^{i(k-n)\psi},$$

$$\overset{2}{\Omega}{}_2^- = \frac{\lambda}{|\lambda|}\left(\frac{R}{\rho}\right)^{\nu+\frac{n}{2}} \sum_{k=n-m}^{\infty} \frac{K_{p_k-1}\left(\frac{|\lambda|}{2\nu}\rho^{-2\nu}\right)}{K_{p_k}\left(\frac{|\lambda|}{2\nu}R^{-2\nu}\right)}\left(\frac{R}{z}\right)^{k-n} e^{i(k-n)\psi},$$

We also introduce the operators $\overset{p}{S}{}_G^-(p = 1, 2)$:

$$\overset{p}{S}{}_G^- f = -\frac{z^{-m}}{\pi}\iint\limits_{G+}\left[\frac{\overset{p}{\Omega}{}_1^-(z,\zeta)}{\zeta}f(\zeta) + \frac{\overset{p}{\Omega}{}_1^-(z,\zeta)}{\bar{\zeta}}\overline{f(\zeta)}\right] d\xi\, d\eta.$$

It is easy to prove that $\overset{1}{S}{}_G^- f$ with $m \geq n_O$ and $\overset{2}{S}{}_G^- f$ with $m < n_O$ are analytic functions in the domain $|z| \geq R$; $\overset{1}{S}{}_G^- f$ having (at infinity) a zero of order $m - n_O + 1$, if n is odd, and order $m - n_O$, if n is even; $\overset{2}{S}{}_G^- f$ is bounded.

9.2 Now we pass over to the solution of the problem. We rewrite the main integral equation in appropriate notation:

$$w^+(z) = \Phi^+(z) + S_G F + P_G \overline{w^+}. \tag{9.3}$$

Let us represent $\Phi^+(z)$ in the form

$$\Phi^+(z) = \Phi_O^+(z) + S_G^+ F + P_G^+ \overline{w^+},$$

where $\Phi_O^+(z)$ is a new unknown solution of equation (8.8) and

$$P_G^+ \overline{w^+} = S_G^+ \left(\frac{b(\zeta) - b(0)}{2\rho^{1+2\nu}} \mathrm{e}^{\mathrm{i}(n+1)\psi} \overline{w^+(\zeta)} \right).$$

Introducing $\Phi^+(z)$ into (9.3) and denoting $\breve{S}_G = S_G + S_G^+$ and $\breve{P}_G = P_G + P_G^+$ we get

$$w^+(z) = \Phi_O^+(z) + \breve{S}_G F + \breve{P}_G \overline{w^+}, \quad z \in G^+. \tag{9.4}$$

Using this $\Phi^-(z)$ is expressed as follows:

$$\Phi^-(z) = \Phi_O^-(z) + \overset{p}{S}{}_G^- F + \overset{p}{P}{}_G^- \overline{w^+}, \quad z \in G^-, \quad (p = 1, 2), \tag{9.5p}$$

where $\Phi_O^-(z)$ is a new analytic function in the domain G^- and

$$\overset{p}{P}{}_G^- \overline{w^+} = \overset{p}{S}{}_G^- \left(\frac{b(\zeta) - b(0)}{2\rho^{1+2\nu}} \mathrm{e}^{\mathrm{i}(n+1)\psi} \overline{w^+(\zeta)} \right).$$

The further argument depends on the sign of $m - n_O$.

9.3 The case $m \geq n_O$ Substituting expressions $w^+(z)$ and $\Phi^-(z)$ from (9.4), $(9.5)_1$) into (9.2) and taking the equality

$$\breve{S}_G F + \breve{P}_G \overline{w^+} = z^m (S_G^- F + \overset{1}{P}{}_G^- \overline{w^+}), \quad z \in \Gamma,$$

into account we get

$$\Phi_O^+(z) = z^m \Phi_O^-(z) + g(z), \quad z \in \Gamma. \tag{9.6}$$

Consequently, the functions $w_O^+(\zeta)$ and $\Phi_O^-(\zeta)$ will satisfy boundary condition (9.2), only if $\Phi_O^+(z)$ and $\Phi_O^-(z)$ are solutions of problem (9.6). The solution of the latter has been obtained in subsection 6.3. Introducing $\Phi_O^+(z)$ and $\Phi_O^-(z)$ into (9.4) and

(9.5$_1$), we discover that the conjugation problem is reduced to finding $w_O^+(z)$ from integral equation (9.4) and defining $\Phi_O^-(z)$ from (9.5$_1$).

Theorem 9.1 ($m \geq n_O$) *Let n be an odd number and* $\|\check{P}_G\|_C < 1$. *Then the homogeneous problem* ($F = g = 0$) *has* $2(m - n_O + 1)$ *linearly independent solutions and the inhomogeneous problem is solvable for any* $F(z)$ *and* $g(z)$.

All statements of this theorem are valid with even n, except the number of linearly independent solutions of the homogeneous problem, which is equal to $2(m - n_O) + 1$.

9.4 The case $m < n_O$ We write down the necessary solvability conditions, which restrict the function $w^+(z)$:

$$T_k^*\left[F(\zeta) + \frac{b(\zeta) - b(0)}{2\rho^{1+2\nu}} e^{in\psi}\overline{w^+(\zeta)}\right] = g_{n-k} - \frac{\lambda}{|\lambda|} \frac{K_{p_k-1}\left(\frac{|\lambda|}{2\nu}R^{-2\nu}\right)}{K_{p_k}\left(\frac{|\lambda|}{2\nu}R^{-2\nu}\right)} \overline{g_k}, \tag{9.7}$$

$$k = n_O, n_O + 1, \ldots, n - m - 1.$$

where g_k, g_{n-k} are Fourier coefficients of the function $g(z), z \in \Gamma$, and

$$T_k^*(f(\zeta)) = \frac{1}{\pi K_{p_k}\left(\frac{|\lambda|}{2\nu}R^{-2\nu}\right)} \iint\limits_{G+} \left(\frac{R}{\rho}\right)^{\nu+\frac{n}{2}} \left[K_{p_k}\left(\frac{|\lambda|}{2\nu}\rho^{-2\nu}\right) e^{i(k-n)\psi} \frac{f(\zeta)}{\zeta}\right.$$

$$\left. - \frac{\lambda}{|\lambda|} K_{p_k}\left(\frac{|\lambda|}{2\nu}\rho^{-2\nu}\right) e^{ik\psi} \overline{\frac{f(\zeta)}{\zeta}}\right] d\xi\, d\eta.$$

The equality (9.7) is obtained from solvability condition (6.26) if in place of $g(z)$ we substitute $g(z) - \check{S}_G F + \check{P}_G \overline{w^+}$.

Introducing expressions (9.4)*and*(9.5$_2$) for $w^+(z)$ and $\Phi^-(z)$ into (9.4) and using (9.7), we arrive at the relation:

$$\Phi_O^+(z) = z^m \Phi_O^-(z) + g^*(z), \quad z \in \Gamma, \tag{9.8}$$

$$g^*(z) = g(z) - \sum_{k=n_O}^{n-m-1} \left[g_{n-k} - \frac{\lambda}{|\lambda|} \frac{K_{p_k-1}\left(\frac{|\lambda|}{2\nu}R^{-2\nu}\right)}{K_{p_k}\left(\frac{|\lambda|}{2\nu}R^{-2\nu}\right)} \overline{g_k}\right] e^{i(n-k)\varphi}.$$

The function $g^*(z)$ automatically satisfies the solvability conditions which are given in Theorem 6.3 and problem (9.8) has unique solution. Therefore the solution to the conjugation problem is reduced to finding $w^+(z)$ from integral equation (9.4), fulfilling the supplementary conditions (9.7) and determining the analytic function $\Phi^-(z)$ by formula (9.5_2).

Theorem 9.2 $(m < n_O)$ *If* $\|\check{P}_G\|_C < 1$, *then the homogeneous problem has only the trivial solution and the inhomogeneous problem is solvable if and only if* $F(z)$ *and* $g(z)$ *satisfy* $2(n_O - m - 1)$ *real solvability conditions with odd* n *and* $2(n_O - m) - 1$ *conditions with even* n.

10 Unsolved problems

10.1 The investigation carried out in this chapter contains some incomplete elements. They arise when we deal with the general integral equation

$$w(Z) = \Phi(Z) + S_G f + P_G \bar{w}, \quad z \in G,$$

and with its modifications, see (8.13), (8.19), (8.22), (9.4). In this case the complete solvability picture was formulated only under the conditions of the compactness of the operators.

In this connection the following problem arises: *it is possible to obtain detailed information on the properties of an operator for obtaining solvability results under less restrictive assumptions?*

10.2 The answer to this question is possibly connected with a proof of the similarity formula

$$w(z) = \Phi(z)\mathrm{e}^{\omega(z)} \tag{10.1}$$

for solutions $w(z)$ and $\Phi(z)$ of the general and model equations:

$$\partial_{\bar{z}} w - \frac{b(z)}{2r^{1+2\nu}}\mathrm{e}^{\mathrm{i}(n+1)\varphi}\bar{w} = 0, \tag{10.2}$$

$$\partial_{\bar{z}} \Phi - \frac{b(0)}{2r^{1+2\nu}}\mathrm{e}^{\mathrm{i}(n+1)\varphi}\bar{\Phi} = 0. \tag{10.3}$$

Differentiating equality (10.1) with respect to $\bar{z}$ and then excluding $\partial_{\bar{z}} w$ and $\partial_{\bar{z}} \Phi$ by means of (10.2) and (10.3), we get

$$\partial_{\bar{z}} \omega + \frac{b(0)}{2r^{1+2\nu}}\mathrm{e}^{\mathrm{i}(n+1)\varphi}\frac{\bar{w}}{w}\mathrm{e}^{2\mathrm{i}\,\mathrm{Im}\omega} = \frac{b(z)}{2r^{1+2\nu}}\frac{\bar{w}}{w}\mathrm{e}^{\mathrm{i}(n+1)\varphi}. \tag{10.4}$$

Now the essence of the problem consists of the following: *is it possible for an arbitrary solution* $w(z) \in C(\bar{G}) \cap C_{\bar{z}}(G-0)$ *of equation (10.2) to yield a solution* $\omega(z)$ *equation* (10.4) *continuous in domain* G?

If such a result were established, then at the next stage it would be necessary to analyse the variety of solutions $w(z)$ of equation (10.2) on the whole complex plane. Then we should get an exhaustive solvability picture of the general and modified integral equations.

10.3 In relation to general equation (10.2) the concrete form of a model equation will evidently depend on the properties of the coefficient $b(z)$. If, for instance, $b(z)$ is a sufficiently regular function in a neighbourhood of $z = 0$, then

$$b(z) = b_m(z, \bar{z}) + O(|z|^{m+1}), \quad z \to 0,$$

where

$$b_m(z, \bar{z}) = \sum_{p=O}^{m} \sum_{k=O}^{p} \beta_{p-k,k} z^{p-k} \bar{z}^k$$

and

$$\beta_{p-k,k} = \frac{\partial^p b(0)}{\partial z^{p-k} \bar{z}^k};$$

in addition m is as follows

$$\frac{b(z) - b_m(z, \bar{z})}{2r^{1+2\nu}} \in L_p(G), \quad p > 2.$$

This condition is fulfilled when m is equal to the integer part of 2ν, i.e. $m = [2\nu]$. In this case, as we can assume, the equation modelling (10.2) will be the following:

$$\partial_{\bar{z}} \Phi - \frac{b_m(z, \bar{z})}{2r^{1+2\nu}} e^{i(n+1)\varphi} \bar{\Phi} = 0. \qquad (10.5)$$

Hence one can see that in the present chapter we restrict ourselves to considering only the partial case (10.3) from the variety (10.5) of the model equations. In this connection the development of equation (10.5) is up–to–date.

Chapter 6
Infinitesimal bendings of surfaces of positive curvature with a flat point

In the present chapter the object of study is a surface given in a rectangular Cartesian coordinate system by the equation

$$z = (x^2 + y^2)^{\frac{n}{2}} f(x,y), \quad (x,y) \in D,$$

where $n > 2$ is any real number. It is assumed that D is a simply connected domain, bounded by a Ljapunov contour, and the point $(0,0) \in D$. In addition $f(x,y) \in C^3(D), f(0,0) > 0$ and for all points of D other than (0,0) the Gaussian curvature of the surface is positive.

The point (0,0) is a flat point on the surface. At this point not only the Gaussian curvature but also the curvatures of all normal cross–sections and hence all coefficients of the second quadratic form vanish. Also at this point the surface with its tangent plane has a contact order greater than 1.

The direct utilization of the apparatus of generalized analytic functions in studying infinitesimal bendings of these objects proves to be impossible. The cause is that at its basis there is an essential restriction on the Gaussian curvature $K \geq k_O > 0$, but on the surface of interest it is broken at the flat point, where $K = 0$.

Therefore in the development of complex analysis we have to prove again the existence of conjugate isometric coordinates, to study their asymptotic behaviour in a neighbourhood of a flat point and to find a singularity in the coefficients of generazlied Cauchy–Riemann systems.

For the surface under study we will show that its infinitesimal bendings are described by the singular integral equation of ch. 3. In the case of $f(x,y) \equiv 1$, we shall deal with a model surface and its deformation will be characterized by the model equation of ch. 2.

The main results of the present chapter have been published earlier by the author [38, 53, 62, 64, 66–68]. For a more detailed study of bendings problem with a flat point see papers [8, 15–22, 25, 31, 45, 49].

1 Equations of infinitesimal bendings of a surface with positive curvature

As is known, two quadratic differential forms (one of them being positively defined), satisfying the Gauss and Peterson–Codazzi equations, determine precisely, up to motions of the space, the surface, for which the above forms are basic; see the Bonnet theorem [28]. Therefore the problem of finding the surfaces, which are infinitesimal bendings of a given surface, can be reduced to investigating solutions of the Gauss and Peterson–Codazzi equations where the coefficients of the second quadratic form are unknown functions and the coefficients of the first quadratic form are considered to be given.

If one studies the deformation of some surface S, then it is convenient to examine the variations $\delta L, \delta M, \delta N$ of the coefficients L, M, N of its second quadratic form (the corresponding coefficients of a deformed surface S_ε will be $L+\varepsilon\delta L, M+\varepsilon\delta M, N+\varepsilon\delta N$ where ε is a small parameter) as unknown functions. The equations for $\delta L, \delta M, \delta N$ issue from the Gauss and Peterson–Codazzi equations by means of their variations, see [69], ch. 5, §7:

$$L\delta N + N\delta L - 2M\delta M = 0,$$

$$\frac{\partial}{\partial x}(\delta M) - \frac{\partial}{\partial y}(\delta L) + \Gamma^1_{12}\delta L - (\Gamma^1_{12} - \Gamma^2_{12})\delta M - \Gamma^2_{11}\delta N = 0,$$

$$\frac{\partial}{\partial y}(\delta M) - \frac{\partial}{\partial x}(\delta N) - \Gamma^1_{12}\delta L - (\Gamma^2_{22} - \Gamma^1_{12})\delta M - \Gamma^2_{12}\delta N = 0,$$

where x, y are interior coordinates on S and $\Gamma^\alpha_{\beta\gamma}$ are the Christoffel symbols of the second kind.

Let us assume that a sufficiently regular surface S has a positive curvature. Then there exists a conjugate isometric system of coordinates u, v such that the coefficients of the second form satisfy the relations: $L(u,v) = N(u,v)$ and $M(u,v) = 0$, [69]. In this system of coordinates the equations for $\delta L(u,v), \delta M(u,v), \delta N(u,v)$ will have the form:

$$\delta N = -\delta L,$$

$$\frac{\partial}{\partial u}(\delta M) - \frac{\partial}{\partial v}(\delta L) - (\Gamma^1_{11} - \Gamma^2_{12})\delta M + (\Gamma^1_{12} + \Gamma^2_{12})\delta L = 0,$$

$$\frac{\partial}{\partial v}(\delta M) - \frac{\partial}{\partial u}(\delta L) - (\Gamma^2_{22} - \Gamma^1_{12})\delta M - (\Gamma^1_{22}) + \Gamma^2_{12})\delta L = 0.$$

Using the notation $w = u + \mathrm{i}v, \mathrm{i}^2 = -1$, and introducing the complex function

$$U(w) = K^{1/4}(w)(\delta M + \mathrm{i}\delta L), \tag{1.1}$$

where $K(w)$ is the Gaussian curvature on the surface S, we write the preceding system in the following form:

$$\partial_{\bar{w}} U + \overline{B(w)} \cdot \bar{U} = 0. \tag{1.2}$$

Here

$$\partial_{\bar{w}} = \frac{1}{2}\left(\frac{\partial}{\partial x} + \mathrm{i}\frac{\partial}{\partial y}\right)$$

and the bar over a letter denotes complex conjugation. In addition

$$B(w) = -\mathrm{e}^{\mathrm{i}\psi(w)}\partial_{\bar{w}}\left(\operatorname{Arch}\frac{H}{\sqrt{K}}\right) - \frac{\partial_w K}{4K}\mathrm{e}^{2\mathrm{i}\psi(w)}, \tag{1.3}$$

where

$$\mathrm{e}^{\mathrm{i}\psi(w)} = \frac{a^-(w)\sqrt{K(w)}}{2\sqrt{a(w)\mathcal{F}(w)}}. \tag{1.4}$$

Here $H(w)$ is the mean curvature of the surface, $\mathcal{F}(w) = H^2 - K^2$ is the Euler difference, $a(w) = EG - F^2$ is the discriminant of the first fundamental quadratic form (E, F, G are its coefficients), $a^-(w) = E - G + 2\mathrm{i}F$, see [69].

Equation (1.2) is complex notation for the generalized Cauchy–Riemann system. It plays an auxiliary part which is the basis for obtaining equations of infinitesimal bendings of the surface with a flat point.

2 Conjugate isometric system of coordinates on the model surface

Let S be a sufficiently regular surface which has a positive Gaussian curvature everywhere except at some interior point O, being a flat point. The question is if conjugate isometric coordinates on $S_O = S - O$ exist, i.e. on the surface S without the flat point. The answer to this question is not evident. It is not contained in [69] because the important restriction to the Gaussian curvature $K \geq k_O > 0$ (uniformly bounded below) does not occur for S_O. In this connection it is necessary to prove the existence of conjugate isometric coordinates on S_O.

In the present section we examine a surface of a special type that helps us to write out the above coordinates explicitly and to discover the specific character in the solution of the formulated problem for the surface with a flat point.

2.1 Let $Oxyz$ be a rectangular Cartesian coordinates. We consider the rotation surface

$$z = (x^2 + y^2)^{n/2} \tag{2.1}$$

in the neighbourhood of the origin $O(0,0,0)$; here n is any real number, $n > 2$. It is clear that O is the flat point of the surface and the Gaussian curvature around it is positive.

The conjugate isometric system of coordinates (u, v) on surface (2.1) is given by the formulas:

$$u = xr^{\sqrt{n-1}-1}$$

$$v = yr^{\sqrt{n-1}-1}, \quad r^2 = x^2 + y^2. \tag{2.2}$$

In fact, we compute the coefficients L, M, N of the second quadratic form of surface (2.1) in coordinates (x, y), see for example [27]:

$$\begin{aligned} L &= \frac{z_{xx}}{\sqrt{1 + z_x^2 + z_y^2}} \equiv \frac{nr^{n-2}[1 + (n-2)x^2/r^2]}{\sqrt{1 + n^2 r^{2n-2}}}, \\ M &= \frac{z_{xy}}{\sqrt{1 + z_x^2 + z_y^2}} \equiv \frac{n(n-2)r^{n-4}xy}{\sqrt{1 + n^2 r^{2n-2}}}, \\ N &= \frac{z_{yy}}{\sqrt{1 + z_x^2 + z_y^2}} \equiv \frac{nr^{n-2}[1 + (n-2)y^2/r^2]}{\sqrt{1 + n^2 r^{2n-2}}}, \end{aligned} \quad . \tag{2.3}$$

Hence the second quadratic form of the surface is as follows

$$\begin{aligned} \mathrm{II} = &\frac{nr^{n-2}}{\sqrt{1 + n^2 r^{2n-2}}} \left\{ \left[1 + (n-2)\frac{x^2}{r^2} \right] dx^2 + 2(n-2)\frac{xy}{r^2}\, dx\, dy \right. \\ &\left. + \left[1 + (n-2)\frac{y^2}{r^2} \right] dy^2 \right\}. \end{aligned}$$

Calculating, from (2.2),

$$\begin{aligned} du^2 + dv^2 = &r^{2\sqrt{n-1}-2} \left\{ \left[1 + (n-2)\frac{x^2}{r^2} \right] dx^2 + 2(n-2)\frac{xy}{r^2}\, dx\, dy \right. \\ &\left. + \left[1 + (n-2)\frac{y^2}{r^2} \right] dy^2 \right\} \end{aligned}$$

and using the result for transforming the second quadratic form, we establish that the second quadratic form in the variables u, v has indeed the isometric form:

$$\mathrm{II} = \frac{n\rho^{n/\sqrt{n-1}-2}}{\sqrt{1+n^2\rho^{2\sqrt{n-1}}}}(\mathrm{d}u^2 + \mathrm{d}v^2). \tag{2.4}$$

In this expression we employed the notation $\rho^2 = u^2 + v^2$ and the equality

$$\rho = r^{\sqrt{n-1}}, \tag{2.5}$$

which follows from (2.2).

Because of (2.4) the coefficients L, M, N of the second quadratic form of surface (2.1) in u, v coordinates are defined in the following way:

$$L(u,v) = N(u,v) = \frac{n\rho^{n/\sqrt{n-1}-2}}{\sqrt{1+n^2\rho^{2\sqrt{n-1}}}}; \quad M(u,v) = 0. \tag{2.6}$$

It should be noted that the change from x, y coordinates to u, v conjugate isometric coordinates according to (2.2) is the similitude transformation with expansion coefficient equal to $r^{\sqrt{n-1}-1}$.

2.2 Let us pay attention to the specific formula (2.2). The Jacobian transform from the old variables x, y to the new variables u, v

$$\frac{\partial(u,v)}{\partial(x,y)} = \sqrt{n-1}\cdot r^{2\sqrt{n-1}-2}$$

vanishes at the flat point $x = y = 0$, since $\sqrt{n-1} - 1 > 0$ for $n > 2$. In spite of this, formula (2.2) establishes the correspondence between the flat point and the origin $u = v = 0$ of the $u, v-$plane, and vice versa.

Thus, we have succeeded in constructing conjugate isometric coordinates on the surface (2.1), but it is necessary that the Jacobian transform vanishes at the flat point. This means that the initial system of coordinates (x, y) and the new system (u, v) are disparate. We should take this fact into account in future.

2.3 Let us consider the problem of constructing a conjugate isometric parametrization on surface (2.1), arising from the Beltrami equation [69]:

$$\partial_{\bar{z}} w - q(z)\partial_z w = 0, \quad z \in D, \tag{2.7}$$

where $z = x + iy, w = u + iv, i^2 = -1, \partial_{\bar z} = \frac{1}{2}(\partial_x + i\partial_y), \partial_z = \frac{1}{2}(\partial_x - i\partial_y)$ and

$$q(z) = \frac{L - N + 2iM}{L + N + 2\sqrt{LN - M^2}}. \tag{2.8}$$

It is known that the proof of the existence of conjugate isometric coordinates on a surface is equivalent to finding the so–called homeomorphic solutions $w = w(z)$ of equation (2.7). These solutions establish a one–to–one mutually continuous correspondence of the domain D of the plane $z = x + iy$ into the domain Δ of the plane $w = u + iv$.

Substituting the expressions L, M, N from (2.3) into (2.8) we get

$$q(z) = q_O e^{2i\varphi},$$

where $\varphi = \arg z$ and

$$q_O = \frac{n-2}{n + 2\sqrt{n-1}} < 1.$$

In any domain, not containing the flat point, the modulus of the coefficient of the Beltrami equation is strictly less than 1 and is a sufficiently regular function. These conditions are sufficient to construct a conjugate isometric parametrization on a corresponding part of the surface.

For the equation

$$\partial_{\bar z} w - q_O e^{2i\varphi} \partial_z w = 0 \tag{2.9}$$

one of its homeomorphic solutions is given by formula (2.2). In complex notation this has the form

$$w = |z|^{\sqrt{n-1}} e^{i\varphi}. \tag{2.10}$$

It remains to consider the homeomorphism in the domain containing the flat point. If we consider (2.9) in this domain then its coefficient, having the factor $e^{2i\varphi}$, will have a bounded break at $z = 0$, i.e. at the flat point. This is one of the features that characterizes the flat point.

3 Conjugate isometric system of coordinates on the general surface

Let us examine the surface

$$z = (x^2 + y^2)^{\frac{n}{2}} f(x, y), \quad (x, y) \in D. \tag{3.1}$$

We assume that all those conditions which were formulated in the introduction to the present chapter are true for (3.1).

We calculate the coefficients E, F, G and L, M, N of the first and second quadratic forms of the surface

$$\left.\begin{aligned} E &= 1 + (r^n f_x + nxr^{n-2} f)^2, \\ F &= (r^n f_x + nxr^{n-2} f)(r^n f_y + nyr^{n-2} f), \\ G &= 1 + (r^n f_y + nyr^{n-2} f)^2; \end{aligned}\right\} \tag{3.2}$$

$$\begin{aligned} L &= \frac{r^{n-2}}{\sqrt{a}}\{[1 + (n-2)\frac{x^2}{r^2}]nf + 2nxf_x + r^2 f_{xx}\}, \\ M &= \frac{r^{n-2}}{\sqrt{a}}\{n(n-2)\frac{xy}{r^2} f + n(xf_y + yf_x) + r^2 f_{xy}\}, \\ N &= \frac{r^{n-2}}{\sqrt{a}}\{[1 + (n-2)\frac{y^2}{r^2}]nf + 2nyf_y + r^2 f_{yy}\}. \end{aligned} \tag{3.3}$$

In formulas (3.3) we used the notation $a = EG - F^2$.

3.1 Let us proceed to the construction of the conjugate isometric system of coordinates (u, v) on the surface (3.1). The search for such coordinates is reduced to proving the existence of homeomorphic solutions of the Beltrami equation (2.7) with the coefficient $q(z)$ expressed by formula (2.8). Substituting the expressions for L, M, N from (3.3) into (2.8) we get

$$q(z) = \frac{n(n-2) f e^{2i\varphi} + 4nzf_{\bar z} + 4r^2 f_{\bar z \bar z}}{n^2 f + 2nrf_r + r^2(f_{xx} + f_{yy}) + 2n\sqrt{n-1} f\sqrt{1+\Pi}},$$

where $\varphi = \arg z$; $f_{\bar z}$ and $f_{\bar z \bar z}$ are the partial derivatives of first and second order with respect to $\bar z$ and

$$\begin{aligned} \Pi = &\frac{1}{n^2(n-1)f^2}\Big\{2n^2 rff_r + nff_{xx}[r^2 + (n-2)y^2] + nff_{yy}[r^2 + (n-2)x^2] \\ &+ 2nr^2[yf_y f_{xx} + xf_x f_{yy}] + r^4(f_{xx}f_{yy} - f_{xy}^2) - n^2(xf_y + yf_x)^2 \\ &- 2n(n-2)xyff_{xy} - 2nr^2(xf_y + yf_x)f_{xy}\Big\}. \end{aligned}$$

Let us express $q(z)$ in the form

$$q(z) = q_O e^{2i\varphi} + q^*(z), \tag{3.4}$$

$$q_O = \frac{n-2}{n+2\sqrt{n-1}} < 1. \tag{3.5}$$

The concrete form of the function $q^*(z)$ is not considered necessary in future. It will be sufficient to know that $q^*(z) \in C^1$ for $|z| > 0$ and $q^*(z) = O(|z|)$, as $z \to 0$. We also indicate that $|q(z)| < 1$ for all $z \in D$. The validity of this inequality at $z = 0$ issues obviously from (3.4), (3.5). It is also true for other values of z since surface (3.1) outside the flat point has a positive curvature.

Thus, in analysing the problem of the existence of a conjugate isometric para metrization on surface (3.1), we deal with the Beltrami equation, in which the modulus of $q(z)$ is strictly less than 1 in the whole domain D and $q(z)$ belongs to the class C^1 outside $z = 0$ and has a bounded break at $z = 0$; see (3.4). But homeomorphic solutions of the Beltrami equation exist even for a wider class of coefficients, i.e. for bounded ($|q| < 1$) and measurable functions. Therefore, the problem of the existence of conjugate isometric coordinates on surface (3.1) has a positive solution.

Later we shall need to know in addition the asymptotic behaviour of the homeomorphism and its partial derivatives as $z \to 0$. The study of this question is dealt with in the next subsection.

3.2 We use conjugate isometric coordinates of surface (2.1) as intermediate interior coordinates on surface (3.1). Let us also agree to use the new variables $\xi, \eta (\zeta = \xi + i\eta)$ instead of the previous notation $u, v (w = u + iv)$. Then formula (2.10) will be written down as follows:

$$\zeta = |z|^{\sqrt{n-1}} e^{i\varphi} \tag{3.6}$$

We turn to equation (2.7), in which $q(z)$ is defined by equality (3.4). Replacing the variables by formula (3.6), we get

$$\partial_{\bar{\zeta}} w - G(\zeta)\partial_{\zeta} w = 0, \quad \zeta \in D_\zeta \tag{3.7}$$

where D_ζ is the image of domain D by the mapping (3.6) and

$$G(\zeta) = \frac{q^*(z)}{1 - q_O^2 - q_O q^*(z) e^{-2i\varphi}}.$$

By means of (3.6), the equality inverse to (3.6)

$$z = |\zeta|^{1/\sqrt{n-1}} e^{i\varphi} \tag{3.8}$$

and the properties of $q^*(z)$ it is easy to state the following properties of $G(\zeta) : G(\zeta) \in C^1$ outside $\zeta = 0$ and $G(\zeta) = O(|\zeta|^{1/\sqrt{n-1}}), \zeta \to 0$. From the last relation it follows that $|G(\zeta)| \leq G_O < 1$ at least in a neighbourhood of the point $\zeta = 0$. Without loss of generality we shall believe that this inequality is fulfilled for $\zeta \in \overline{D_\zeta}$.

Thus, in the variable ζ we have arrived at a case of the Beltrami equation which had been well studied. On the basis of results in [69], ch. 2, §4, we can convince ourselves that the next lemma is reasonable.

Lemma 3.1 *Equation (3.7) has a homeomorphic solution of class $C^1_\alpha, a = 1/\sqrt{n-1}$, by means of which the domain D_ζ is mapped onto the disk $|w| < R$ while the point $\zeta = 0$ corresponds to $w = 0$ and*

$$w = \zeta + O(|\zeta|^{1+1/\sqrt{n-1}}), \qquad \partial_\zeta w = 1 + O(|\zeta|^{1/\sqrt{n-1}}),$$

$$\partial_{\bar\zeta} w = O(|\zeta|^{1/\sqrt{n-1}}), \quad \zeta \to 0. \tag{3.9}$$

3.3 Now considering the function $w(\zeta(z))$, where $\zeta(z)$ is given by (3.6), and $w(\zeta)$ is taken from Lemma 3.1, we get the homeomorphism of equation (2.7). The asymptotic behaviour of the homeomorphism and its partial derivatives as $z \to 0$ is not hard to study, since we have the explicit dependence of ζ on z and formulas (3.9). Since these results will be necessary in §4, we formulate them as a separate lemma.

Lemma 3.2 *The Beltrami equation*

$$\partial_{\bar z} w - [q_O e^{2i\varphi} + q^*(z)]\partial_z w = 0, \quad z \in D, \tag{3.10}$$

$$q_O = \frac{n-2}{n+2\sqrt{n-1}}, \quad n > 2,$$

has a homeomorphic solution $w(z)$, whose partial derivatives outside $z = 0$ satisfy the Holder condition with exponent α arbitrarily close to 1. By means of this homeomorphism D is mapped onto the disk $|w| < R$ in such a way that the point $z = 0$ corresponds to $w = 0$ and

$$w = |z|^{\sqrt{n-1}} e^{i\varphi} + O(|z|^{\sqrt{n-1}+1}), \tag{3.11}$$

$$\partial_z w = \frac{\sqrt{n-1}+1}{2} |z|^{\sqrt{n-1}} + O(|z|^{\sqrt{n-1}+1/\sqrt{n-1}-1}), \quad z \to 0. \tag{3.12}$$

The inverse function $z = z(w)$ is such that $\partial_{\bar{w}} z$ and $\partial_w z$ satisfy the Holder condition for $|w| > 0$, and

$$\partial_{\bar{w}} z + (q_O e^{2i\varphi} + q^*(z))\overline{\partial_w z} = 0, \quad |w| < R, \tag{3.13}$$

and the following relations hold:

$$z = |w|^{1/\sqrt{n-1}} e^{i\varphi} + O(|w|^{2/\sqrt{n-1}}), \tag{3.14}$$

$$\partial_w z = \frac{\sqrt{n-1}+1}{2\sqrt{n-1}} |w|^{1/\sqrt{n-1}-1} \left\{ 1 + O(|w|^{1/(n-1)}) \right\}, \quad w \to 0. \tag{3.15}$$

In fact, (3.11) and (3.12) follow from (3.9) and (3.6). To deduce (3.12) one uses the identity

$$\partial_z w \equiv \partial_\zeta w \cdot \partial_z \zeta + \partial_{\bar{\zeta}} w \cdot \overline{\partial_{\bar{z}} \zeta}.$$

With the help of this identity and the corresponding identity for $\partial_{\bar{z}} w$ one proves also that $\partial_z w$ and $\partial_{\bar{z}} w$ satisfy the Holder condition with an exponent arbitrarily close to 1, outside $z = 0$.

We now consider the inverse function $z = z(w)$. Solving the first relation in (3.9) for ζ, we get

$$\zeta = w + O(|w|^{1+1/\sqrt{n-1}}), \quad w \to 0. \tag{3.16}$$

Substituting (3.16) in (3.8), we get (3.14).

Further we shall make use of the formulas

$$\partial_w z = \frac{1}{J} \overline{\partial_z w}, \quad \partial_{\bar{w}} z = -\frac{1}{J} \partial_{\bar{z}} w, \tag{3.17}$$

where $J = |\partial_z w|^2 - |\partial_{\bar{z}} w|^2$ is the Jacobian transformation $w = w(z)$. From (3.17), taking account of (3.10), it follows that $z(w)$ satisfies (3.13).

The validity of (3.15) is established from the first of formulas (3.17) by the following chain of identities, based on relations (3.10) and (3.12):

$$\begin{aligned} \partial_w z &= \frac{\overline{\partial_z w}}{|\partial_z w|^2 - |\partial_{\bar{z}} w|^2} \equiv \frac{\overline{\partial_z w}}{(1 - |q_O e^{2i\varphi} + q^*(z)|^2)|\partial_z w|^2} \\ &\equiv \frac{(\partial_z w)^{-1}}{1 - q_O^2 + O(|z|)} \equiv \frac{(\sqrt{n-1}+1)|z|^{1-\sqrt{n-1}}[1 + O(|z|^{1/\sqrt{n-1}}]}{2\sqrt{n-1}[1 + O(|z|)]} \\ &\equiv \frac{\sqrt{n-1}+1}{2\sqrt{n-1}} |z|^{1-1/\sqrt{n-1}} [1 + O(|z|^{1/\sqrt{n-1}})], \quad z \to 0. \end{aligned}$$

If we now replace z by w from (3.14), we get (3.15).

The proof of Lemma 3.2 is complete.

Remark If one assumes $f(x,y) \in C^m(\bar{D}), m \geq 3$, then any homeomorphism $w = w(z)$ of equation (3.10) will belong to the class C_a^{m-2} (a is a positive number arbitrarily near to 1) outside $z = O$. This remark is an obvious consequence of the high regularity of the auxiliary homeomorphism (3.6) outside $z = O$ and Theorem 2.12 from [69].

4 Equations of infinitesimal bendings of the general surface

We shall prove that infinitesimal bendings of surface (3.1) are described by the equation

$$\partial_{\bar{w}} U^* - \frac{b(w)}{2\bar{w}} \overline{U^*} = 0, \tag{4.1}$$

where

$$U^*(w) = w^2 K^{1/4}(w)(\delta M + i\delta L), \tag{4.2}$$

$$b(0) = \frac{n-2}{2\sqrt{n-1}}. \tag{4.3}$$

As for surface (2.1) here we have $b(w) \equiv b(0)$.

Our proof relies on equation (1.2). We should evidently study the properties of the coefficient $B(w)$ for surface (3.1). To this end we compute the geometric quantities which occur in defining $B(w)$ by (1.3). First we do this in the old system of coordinates $z = x + iy$; then applying the formulas of transformation of geometric values in passing from one set of variables to another, we get their expressions in conjugate isometric coordinates $w = u + iv$.

4.1 According to (3.2) we have

$$\begin{aligned} a^-(z) &= E - G + 2iF \equiv \gamma^-(z), \\ a(z) &= EG - F^2 \equiv 1 + \gamma(z). \end{aligned} \tag{4.4}$$

We do not give the concrete form of the expressions for $\gamma^-(z)$ and $\gamma(z)$ since we shall not need them in future. It will suffice to know that $\gamma^-(z), \gamma(z) \in C^2(\overline{D})$ and $\gamma^-(z), \gamma(z) = O(r^{2n-2}), z \to O$. We shall need the quantity

$$a^+(z) = E + G \equiv 2 + \gamma^+(z), \tag{4.5}$$

where $\gamma^+(z) \in C^2(\overline{D})$ and $\gamma^+(z) = O(r^{2n-2}), z \to 0$.

Further, we have

$$
\begin{aligned}
K(z) &= \frac{LM - M^2}{EG - F^2} \equiv \frac{r^{2n-4}}{a^2}\{n^2(n-1)f^2 + k(z)\},\\
H(z) &= \frac{LG - 2MF + NG}{2a} \equiv \frac{r^{n-2}}{2a^{3/2}}\{n^2 f + h(z)\},\\
&\frac{H(z)}{\sqrt{K(z)}} \equiv \frac{n}{2\sqrt{n-1}}\{1 + l(z)\},\\
&\mathcal{F}(z) = H^2 - K \equiv \frac{n^2(n-2)^2}{4} r^{2n-4} f^2\{1 + e(z)\},\\
&\frac{K(z)}{\mathcal{F}(z)} \equiv \frac{4(n-1)}{(n-2)^2}\{1 + g(z)\},
\end{aligned}
\tag{4.6}
$$

where $k(z), h(z), l(z), e(z), g(z) \in C^1(\overline{D})$ and for $z = 0$ have zeros of the first order.

4.2 Since the Gaussian curvature K and the mean curvature H are geometric scalars (see for example [27]), their expressions in conjugate isometric coordinates w are obtained from the first two relations (4.6) by the formulas $K(w) = K[z(w)], H(w) = H[z(w)]$. In particular, takaing (3.14) into account, we have

$$K(w) = O(|w|^{2(n-2)/\sqrt{n-1}}), \quad w \to 0. \tag{4.7}$$

The situation is different with the quantities $a(z), a^+(z)$ and $a^-(z)$. Their values depend essentially on the choice of coordinates. In fact, if $r = r(x, y)$ is the equation of the surface S in vector form, then

$$a^+(z) = 4(\partial_z r \cdot \partial_{\bar z} r),$$

$$a^-(z) = 4(\partial_{\bar z} r)^2; \qquad 4a(z) = [a^+(z)]^2 - [a^-(z)|^2;$$

see [69], §2, ch. 2. On the basis of these equalities, we get

$$a^-(w) = \overline{a^-(z)}(\partial_{\bar w} z)^2 + 2a^+(z)\partial_{\bar w} z\overline{\partial_w z} + a^-(z)\overline{(\partial_w z)}^2,$$

$$a(w) = a(z)\{|\partial_w z|^2 - |\partial_{\bar w} z|^2\}^2. \tag{4.8}$$

Starting from (1.4), we can now establish some properties of the function $\exp\{i\psi(w)\}$. On the basis of (4.8), the fifth of (4.6) and Lemma 3.2, we get

$$e^{i\psi(w)} = -\frac{w}{\bar w}(1 + p(w)), \tag{4.9}$$

where for $|w| > 0, p(w)$ satisfies the Holder condition and

$$p(w) = O(|w|^{1/\sqrt{n-1}}), \quad w \to 0.$$

It is perfectly obvious that $|1 + p(w)| \equiv 1$.

Let us study another factor which is part of (1.3):

$$\begin{aligned} \partial_{\bar{w}}\left(\operatorname{Arch}\frac{H}{\sqrt{K}}\right) &= \left(\operatorname{Arch}\frac{H}{\sqrt{K}}\right)_{\bar{z}} \overline{\partial_w z} + \left(\operatorname{Arch}\frac{H}{\sqrt{K}}\right)_z \partial_{\bar{w}} z \\ &\equiv \left[\left(\operatorname{Arch}\frac{H}{\sqrt{K}}\right)_{\bar{z}} - (q_O e^{2i\varphi} + q^*)\left(\operatorname{Arch}\frac{H}{\sqrt{K}}\right)_z\right] \overline{\partial_w z}. \end{aligned}$$

Here we use (3.13) in order to eliminate $\partial_{\bar{w}} z$. Substituting for $H/\sqrt{K}$ its value from the third of formulas (4.6) and applying Lemma 3.2, we obtain

$$\partial_{\bar{w}}\left(\operatorname{Arch}\frac{H}{\sqrt{K}}\right) \in C \quad \text{for } |w| > 0$$

and

$$\partial_{\bar{w}}\left(\operatorname{Arch}\frac{H}{\sqrt{K}}\right) = O(|w|^{1/\sqrt{n-1}-1}), \quad w \to 0. \tag{4.10}$$

Finally we study the ratio $\partial_w K/4K$, which we must also know to determine $B(w)$:

$$\frac{\partial_w K}{K} = \frac{1}{4K}\left[K_z \partial_w z + K_{\bar{z}} \overline{\partial_{\bar{w}} z}\right] \equiv \frac{1}{4K}\left[K_z - (q_O e^{-2i\varphi} + \bar{q}^*) K_{\bar{z}}\right] \partial_w z.$$

Just as before, we used (3.13). Direct differentiation of the Gaussian curvature $K(z)$, which is given by the first formula of (4.6), arrives at the following

$$\frac{K_z}{K} = \frac{n-2}{z} + k_1(z), \qquad \frac{K_{\bar{z}}}{K} = \frac{n-2}{\bar{z}} + k_2(z),$$

where $k_1(z), k_2(z) \in C(D)$. Using these equations and Lemma 3.2, we get that $\partial_w K/4K \in C$ for $|w| > 0$ and

$$\frac{1}{4K}\partial_w K = \frac{n-2}{4\sqrt{n-1}}\frac{1}{w} + O(|w|^{-1/\sqrt{n-1}}), \quad w \to 0. \tag{4.11}$$

Now we have everything necessary for the description of the properties of the coefficient $B(w)$. Since in each summand of (1.3) the factors are continuous for $|w| > 0$, then $B(w)$ is also continuous for $|w| > 0$. Also, with the help of (4.9)–(4.11) one establishes the behaviour of $B(w)$ at the point $w = 0$:

$$B(w) = -\frac{n-2}{4\sqrt{n-1}}\left(\frac{w}{\bar{w}}\right)^2 \frac{1}{w} + O(|w|^{-\delta}), \quad w \to 0.$$

For later use the coefficient $B(w)$ can be conveniently written in the form:

$$B(w) = -\frac{1}{2}\left(\frac{w}{\bar{w}}\right)^2 \cdot \frac{\overline{b(w)}}{w}, \tag{4.12}$$

where $b(w)$ is a continuous function which satisfies the Holder condition at the point $w = 0$ with exponent $1-\delta, \delta = \max(1/\sqrt{n-1}, 1-1/\sqrt{n-1})$ and $b(0)$ is determined by (4.3). Substituting (4.12) into (1.2), we get

$$\partial_{\bar{w}} U - \frac{1}{2}\overline{\left(\frac{w}{\bar{w}}\right)^2 \cdot \frac{b(w)}{\bar{w}}}\bar{U} = 0. \tag{4.13}$$

It remains to introduce the unknown function $U^*(w)$ by equality (4.2), to get (4.1) from (4.13). It should be noted that $U^*(w) = w^2 U(w)$ and (4.1) is given on the disk $|w| < R$, since it is the domain in which the surface S is mapped by means of the conjugate isometric system of coordinates (see Lemma 3.2).

4.3 In contrast to the previous point, for investigating the surface (2.1) it is not sufficient to have only asymptotics of geometric quantities, but it is necessary to get their exact expressions. Omitting the calculations, we write out the formulas of the quantities intended for the determination of $B(w)$:

$$a(w) = \frac{1}{n-1}(1 + n^2|w|^{2\sqrt{n-1}})|w|^{4(1/\sqrt{n-1}-1)};$$

$$a^-(w) = -\frac{n-2-n^2|w|^{2\sqrt{n-1}}}{n-1} \cdot |w|^{2\sqrt{n-1}-4} w^2;$$

$$H(w) = \frac{n^2(1 + n|w|^{2\sqrt{n-1}})|w|^{(n-2)/\sqrt{n-1}}}{2(1 + n^2|w|^{2\sqrt{n-1}})^{3/2}}; \tag{4.14}$$

$$K(w) = \frac{n^2(n-1)|w|^{2(n-2)/\sqrt{n-1}}}{(1 + n^2|w|^{2\sqrt{n-1}})^2};$$

$$e^{i\psi(w)} = -\frac{w^2}{|w|^2}.$$

Substituting these quantities in (1.3), we again have formula (4.12) but only in the situation when

$$b(w) \equiv b(0) = \frac{n-2}{2\sqrt{n-1}}.$$

4.4 To employ in future the generally accepted notation, let us agree to the following: *the conjugate isometric coordinates from now on will be denoted by* $z = x + iy$, *and the unknown function by* $w(z)$, *instead of the previous notation* $w = u + iv$ and $U^*(w)$. The above procedure was used to pass over to ordinary notation. Now we formulate the interrelation of the geomemtric and analytic problems in their final form.

Let S be a simply–connected surface of positive curvature with an isolated flat point, given by equation (3.1). A complex–valued function

$$w(z) = z^2 K^{1/4}(z)(\delta M + \mathrm{i}\delta L), \tag{4.15}$$

continuously differentiable outside $z = O$ and satisfying the equation

$$\partial_{\bar z} w - \frac{b(z)}{2\bar z}\bar w = 0 \tag{4.16}$$

corresponds to each infinitesimal bending of surface (3.1), determined by the variations $\delta L, \delta M, \delta N$ of the coefficients of the second quadratic form. Herer $b(z)$ is continuous and satisfies, at the point $z = 0$, the Holder condition, $b(0)$ being defined by (4.3).

We shall denote by $\Phi(z)$ the complex–valued function $w(z)$ for surface (2.1). It is a solution of the equation

$$\partial_{\bar z} \Phi - \frac{b(0)}{2\bar z}\bar\Phi = 0. \tag{4.17}$$

5 Geometric significance of boundary conditions

Let S be a surface of positive curvature and $z = x + iy$ be a conjugate isometric coordinates on it. We denote the normal curvature and geodesic torsion of some curve L on the surface by k_s and τ_s respectively. According to [69] (see §7 ch.5) for the variations of these quantities we have

$$\delta k_s = -\mathrm{Re}[\mathrm{i}\left(\frac{\mathrm{d}z}{\mathrm{d}s}\right)^2(\delta M + \mathrm{i}\delta L)],$$

$$\delta \tau_s = \mathrm{Re}[\mathrm{i}\frac{\mathrm{d}z}{dl}\frac{\mathrm{d}z}{\mathrm{d}s}(\delta M + \mathrm{i}\delta L)].$$

Thanks to (4.15) these formulas are reduced to the form:

$$\delta k_s = \mathrm{Re}(\gamma_1 w) \quad \text{and} \quad \delta \tau_s = \mathrm{Re}(\gamma_2 w), \tag{5.1}$$

where

$$\gamma_1 = -\mathrm{i}z^{-2}\left(\frac{\mathrm{d}z}{\mathrm{d}s}\right)^2 K^{-1/4}(z) \quad \text{and} \quad \gamma_2 = \mathrm{i}z^{-2}\frac{\mathrm{d}z}{\mathrm{d}l}\frac{\mathrm{d}z}{\mathrm{d}s}K^{-1/4}(z).$$

If L is a boundary curve of a simply–connected piece of the surface, then the increment in the arguments of the functions γ_1 and γ_2 are equal to zero under a single circuit of the curve L.

Let us compose a linear combination of the variation δk_s and $\delta \tau_s$. As a result of (5.1) we get

$$\lambda(s)\delta k_s + \mu(s)\delta\tau_s \equiv \mathrm{Re}(g(z)w(z)). \tag{5.2}$$

Here $\lambda(s), \mu(s)$ are functions, given on L by its length s and $g(z)$ is a function, determined through $\lambda, \mu, \gamma_1, \gamma_2$ and computed in conjugate isometric coordinates.

Equality (5.2) implies that one of the admissible geometric interpretations of the Riemann–Hilbert boundary problem, studied in ch. 2 and 3, is the search for infinitesimal bendings of a surface of positive curvature with a linear combination of δk_s and $\delta \tau_s$ given in advance along the surface edge.

6 A general expression of infinitesimal bendings for the model surface

In the present section and also in the five following ones we leave behind us the methods of complex analysis, and concentrate on obtaining some geometric results useful in our further investigations.

6.1 Let us suppose that in a rectangular Cartesian system of coordinates $Oxyz$ a surface S is given by the equation $z = z(x, y)$. Then its bendings are described by a vector field τ whose components ξ, η, ζ satisfy the system of equations [20]:

$$\begin{gathered} \xi_x + z_x\zeta_x = 0, \quad \eta_y + z_y\zeta_y = 0, \\ \xi_y + \eta_x + z_x\zeta_y + z_y\zeta_x + 0. \end{gathered} \tag{6.1}$$

From (6.1) it follows that the vertical component ζ of τ satisfies the partial differential equation second order:

$$z_{xx}\zeta_{yy} - 2z_{xy}\zeta_{xy} + z_{yy}\zeta_{xx} = 0. \tag{6.2}$$

If $\zeta(x,y)$ is found then the other two components ξ and η will be determined from (6.1) by quadratures [44].

6.2 Let us consider the surface

$$z = r^n \quad (r^2 = x^2 + y^2) \tag{6.3}$$

in a neighbourhood of the point $O(0,0,0)$. Here n is any real number $n > 2$. Equation (6.2) applied to (6.3) takes the form:

$$\left[1 + (n-2)\frac{x^2}{r^2}\right]\zeta_{yy} - 2(n-2)\frac{xy}{r^2}\zeta_{xy} + \left[1 + (n-2)\frac{y^2}{r^2}\right]\zeta_{xx} = 0.$$

Passing to polar coordinates $x = r\cos\varphi, y = r\sin\varphi$, we obtain

$$r^2\zeta_{rr} + (n-1)r\zeta_r + (n-1)\zeta_{\varphi\varphi} = 0. \tag{6.4}$$

Let U be a circular neighbourhood of the point $(0,0)$. Assuming $\zeta(x,y) \in C^2(U)$, we expand $\zeta(x,y)$ in Fourier series:

$$\zeta(x,y) = A_O(r) + \sum_{k=1}^{\infty} A_k(r)\cos k\varphi + B_k(r)\sin k\varphi. \tag{6.5}$$

In the relation to the unknown Fourier coefficients $A_O(r), A_k(r), B_k(r), k = 1,2,\dots,$ from (6.4) we get the system of ordinary differential equations:

$$\begin{aligned} rA''_O + (n-1)A'_O &= 0, \\ r^2A''_k + (n-1)rA'_k - k^2(n-1)A_k &= 0, \\ r^2B''_k + (n-1)rB'_k - k^2(n-1)B_k &= 0, \end{aligned}$$

where the primes over letters denote differentiation with respect to r. Integrating this system and using the boundedness of $A_O(r), A_k(r), B_k(r)$ at the point $r = 0$, we get

$$A_O(r) = a_O, \qquad A_k(r) = a_k r^{\nu_k}, \qquad B_k = b_k r^{\nu_k}, \tag{6.6}$$

where a_O, a_k, b_k are arbitrary constants and

$$\nu_k = \sqrt{\left(\frac{n-2}{2}\right)^2 + k^2(n-1)} - \frac{n-2}{2}. \tag{6.7}$$

By considering an infinitesimal bending without a trivial component, we may assume the bending normalized to give $\zeta(0,0) = \zeta_x(0,0) = \zeta y(0,0) = 0$. Then trivial bendings are just those for which $\zeta(x,y) \equiv 0$.

These normalized conditions are reduced to the relations $A_O = A_1 = B_1 = 0$. Now if we substitute (6.6) in (6.5), then we shall get the variety of all solutions to equation (6.4) from the class C^2 in a neighbourhood of the point $(0,0)$ in the form:

$$\zeta(x,y) = \sum_{k=2}^{\infty}(a_k \cos k\varphi + b_k \sin k\varphi)r^{\nu_k}. \tag{6.8}$$

For $n = 2$, when flattening is absent, $\nu_k = k$ and (6.8) gives an analytic function in the variables x, y.

7 Surfaces of types S_O and S_1

We agree to say that an infinitesimal bending of surface (6.3) belongs to the class C^p or C^A(C^A denotes analyticity) if at least the third component $\zeta(x,y)$ belongs to C^p or C^A.

7.1 Theorem 7.1 *Let $p \geq 2$ be any natural number. Then for any surface (6.3) a neighbourhood of the point* (0,0) *and a non-trivial infinitesimal bending, which belongs to class C^p in this neighbourhood exist.*

In fact, each harmonic of series (6.8) defines an infinitesimal bending of S. The corresponding bending belongs to the class C^p for those values of k, where $\nu_k > p$,

7.2 Let $k \geq 3$ be a fixed integer and N_k be the finite set of real (rational) numbers $n (n > 2)$ defined by

$$n = 2 + 4q\frac{k+q}{k^2 - k - 2q} \quad (q = 1, 2, \ldots, \frac{k^2 - k}{2} - 1). \tag{7.1}$$

Let $N = \bigcup_{k=3}^{\infty} N_k$. Denoting any surface of type (6.3) by the letter S, we shall distinguish surfaces of types S_O and S_1 by either $n \in N$ or $n \notin N$.

Theorem 7.2 *Surfaces of type S_O are non-rigid and surfaces of type S_1 are locally rigid in class C^A.*

Proof Let us assume that for a certain n the surface (6.3) admits an analytic bending in a neighbourhood of (0,0):

$$\zeta(x,y) = \sum_{k=2}^{\infty} \Phi^{(m)}(x,y),$$

where $\Phi^{(m)}(x,y)$ is a homogeneous form of degree m with respect to x and y. Then, in view of the specific formula for the surface (6.3) each of these forms is a solution of equation (6.4). Let us consider one of them and assume $x = r\cos\varphi, y = r\sin\varphi$. We obtain

$$\Phi^{(m)}(x,y) = r^m f_m(\varphi), \tag{7.2}$$

where $f_m(\varphi)$ is a regular $2\pi-$periodic function. Substituting (7.2) in (6.4), we get

$$(n-1)f''_m(\varphi) + (m^2 - 2m + mn)f_m(\varphi) = 0. \tag{7.3}$$

Hence

$$m^2 - 2m + mn = k^2(n-1),$$

where k is an integer. Depending on $n (2 < n < \infty)$ the value of the number k can vary in the limits $m < k^2 < m^2$, and since $m \geq 2$, then $k \geq 3$.

Determining $f_m(\varphi)$ from (7.3) and introducing its expression into (7.2), we get

$$\Phi^{(m)}(x,y) = r^m(a_k \cos k\varphi + b_k \sin k\varphi). \tag{7.4}$$

From (7.4) we find: $m = \nu_k$, where ν_k is given by formula (6.7). Consequently, $\Phi^{(m)}(x,y)$ will have the form

$$\Phi^{(m)}(x,y) = r^{\nu_k}(a_k \cos k\varphi + b_k \sin k\varphi).$$

Thus, a surface of type (6.3) admits an analytic infinitesimal bending if and only if at least one harmonic of a series such as (6.8) is an analytic function. But for this it is necessary and sufficient that for a certain k

$$\nu_k = k + 2q, \tag{7.5}$$

where q is a natural number. Then

$$r^{\nu_k}(a_k \cos k\varphi + b_k \sin k\varphi) = [a_k \mathrm{Re}(x+\mathrm{i}y)^k + b_k \mathrm{Im}(x+\mathrm{i}y)^k](x^2+y^2)^q.$$

Further, from (6.7) we get: $k < \nu_k < k^2$ and from (7.5) it follows that $0 < q < (k^2 - k)/2$. Hence after solving simultaneously (6.7) and (7.5) with respect to n we find (7.1). Theorem 7.2 is proved.

8 Properties of infinitesimal bendings of class C^p for surfaces of type S_1

In the present section the following assertions will be proved.

Theorem 8.1 *For a surface of type S_1 an infinitesimal bending of class $C^p, p \geq 2$, is an arbitrarily small value, characterized by the relation:*

$$\zeta(x,y) = O(r^{\nu_{k_{n,p}+1}}), \quad r \to 0,$$

where

$$\nu_{k_{n,p}+1} = \sqrt{\left(\frac{n-2}{2}\right)^2 + (k_{n,p}+1)^2(n-1)} - \frac{n-2}{2}$$

and $k_{n,p}$ is defined by the inequalities

$$\frac{\sqrt{p(p+n-2)}}{n-1} - 1 < k_{n,p} \leq \frac{\sqrt{p(p+n-2)}}{n-1}. \tag{8.1}$$

Theorem 8.1 admits not only an asymptotic but also a quantitative (estimating) formula. Namely, we have

Theorem 8.2 *For a surface of type S_1 we assume that $\zeta(x,y) \in C^p$ and that on the circle $r = R$ we have $|\zeta| \leq M$. Then for any point $(x = r\cos\varphi, y = r\sin\varphi), r < R$,*

$$|\zeta(x,y)| < \frac{4MR}{R-r}\left(\frac{r}{R}\right)^{\nu_{k_{n,p}+1}}. \tag{8.2}$$

Theorem 8.3 *A surface of type S_1 is locally rigid in class C^∞.*

8.1 Proof of Theorem 8.1 Let the surface be of type S_1. Then (7.5) is impossible. Under this condition if $\zeta(x,y) \in C^p (p > 2)$, then in the expression (6.8) $a_2 = b_2 = \ldots = a_{k_{n,p}} = b_{k_{n,p}} = 0$, where the natural number $k_{n,p}$ is defined by

$$\nu_{k_{n,p}} \leq p < \nu_{k_{n,p}+1}.$$

Solving this with respect to $k_{n,p}$ we have (8.1). Consequently, $\zeta = O(r^{\nu_{k_{n,p}+1}}), r \to 0$.

Remark Under the conditions of Theorem 8.1 it is clear that

$$\zeta_{xx}, \zeta_{xy}, \zeta_{yy} = O(r^{\nu_{k_{n,p}+1}-2}), \quad r \to 0.$$

We shall use this in future.

8.2 Proof of Theorem 8.2 From (6.8) we have

$$R^{\nu_k} a_k = \frac{1}{\pi}\int_0^{2\pi} [\zeta]_{r=R} \cos k\varphi \, d\varphi,$$

$$R^{\nu_k} b_k = \frac{1}{\pi}\int_0^{2\pi} [\zeta]_{r=R} \sin k\varphi \, d\varphi.$$

The conditions of Theorem 8.2 show us that

$$|a_k|, |b_k| < \frac{2M}{R^{\nu_k}}.$$

From this and from $\zeta(x,y) \in C^p$, via (6.8), we have

$$|\zeta(x,y)| < 4M\left(\frac{r}{R}\right)^{\nu_{k_{n,p}+1}} \sum_{k_{n,p}+1}^{\infty} \left(\frac{r}{R}\right)^{\nu_k - \nu_{k_{n,p}+1}}. \tag{8.3}$$

Taking into account the inequality

$$\nu_{k+q} - \nu_k \geq q,$$

we obtain the estimate (8.2) from (8.3).

8.3 Proof of Theorem 8.3 For a circular neighbourhood of the point (0,0) we have $\zeta = 0$ from $\zeta \in C^\infty$ and Theorem 8.1 (or Theorem 8.2). But then $\zeta(x,y) \equiv 0$ in any neighbourhood, since, in view of the elliptic character of the problem outside (0,0), the function $\zeta(x,y)$ is analytic at all points $(x,y) \neq (0,0)$[7].

We note that the results of §§6–8 were obtained in [22].

9 Properties of smallness in deformation of class C^p for the general surface

Let us consider the surface of a more general kind:

$$z = r^n f(x,y), \quad n > 2, \tag{9.1}$$

where $r^2 = x^2 + y^2, f(x,y)$ is a sufficiently regular function in the domain D for which the point (0,0) is interior and $f(0,0) \neq 0$. We assume that (9.1) has a positive curvature outside (0,0).

As in §7, denoting any surface of type (9.1) by S, we shall distinguish surfaces of types S_O and S_1, if $n \in N$ or $n \notin N$, respectively. It is important to note that this separation does not depend on $f(x,y)$.

In the present section for surfaces of type S_1 we extend the result which was stated in the remark to Theorem 8.1.

Theorem 9.1 *For a surface of type S_1 an infinitesimal bending of class $C^p, p \geq 2$, satisfies the relations:*

$$\zeta_{xx}, \zeta_{xy}, \zeta_{yy} = O(r^{p-2}), \quad r \to 0. \tag{9.2}$$

Proof For surface (9.1) equation (6.2) takes the form

$$\left[1 + (n-2)\frac{x^2}{r^2} + 2x\frac{f_x}{f} + r^2\frac{f_{xx}}{nf}\right]\zeta_{yy}$$

$$-2\left[(n-2)\frac{xy}{r^2} + \frac{xf_y + yf_x}{f} + r^2\frac{f_{xy}}{nf}\right]\zeta_{xy}$$

$$+\left[1 + (n-2)\frac{y^2}{r^2} + 2y\frac{f_y}{f} + r^2\frac{f_{yy}}{nf}\right]\zeta_{xx} = 0. \tag{9.3}$$

Since $\zeta(x,y) \in C^p$, then we can expand $\zeta(x,y)$ in a partial Taylor series with a remainder term in a neighbourhood of $(0,0)$:

$$\zeta(x,y) = \sum_{k=0}^{p-1} \Phi^{(k)}(x,y) + R(x,y), \tag{9.4}$$

where

$$\Phi^{(k)}(x,y) = \frac{1}{k!}\sum_{s=O}^{k} \frac{\partial^k \zeta(0,0)}{\partial x^s \partial y^{k-s}} x^s y^{k-s}$$

and $R(x,y) \in C^p$; in addition

$$R_{xx}, R_{xy}, R_{yy} = O(r^{p-2}), \quad r \to 0. \tag{9.5}$$

We prove that $\Phi^{(k)}(x,y) \equiv 0, k = 2, \ldots, p-1$. Then (9.2) will issue from (9.4), (9.5).

Substituting (9.4) into (9.3), we get

$$
\begin{aligned}
&[1+(n-2)\frac{x^2}{r^2}]\Phi^{(2)}_{yy} - 2(n-2)\frac{xy}{r^2}\Phi^{(2)}_{xy} + [1+(n-2)\frac{y^2}{r^2}]\Phi^{(2)}_{xx} \\
&+[2x\frac{f_x}{f} + r^2\frac{f_{xx}}{nf}]\Phi^{(2)}_{yy} - 2[\frac{xf_y+yf_x}{f} + r^2\frac{f_{xy}}{nf}]\Phi^{(2)}_{xy} \\
&+[2y\frac{f_y}{f} + r^2\frac{f_{yy}}{nf}]\Phi^{(2)}_{xx} \\
&+[1+(n-2)\frac{x^2}{r^2} + 2x\frac{f_x}{f} + r^2\frac{f_{xx}}{nf}]\left(\sum_{k=3}^{p-1}\Phi^{(k)}_{yy} + R_{yy}\right) \\
&-2\left[(n-2)\frac{xy}{r^2} + \frac{xf_y+yf_x}{f} + r^2\frac{f_{xy}}{nf}\right]\left(\sum_{k=3}^{p-1}\Phi^{(k)}_{xy} + R_{xy}\right) \\
&+\left[1+(n-2)\frac{y^2}{r^2} + 2y\frac{f_y}{f} + r^2\frac{f_{yy}}{nf}\right]\left(\sum_{k=3}^{p-1}\Phi^{(k)}_{xx} + R_{xx}\right) = 0.
\end{aligned}
\tag{9.6}
$$

The first three summands of this equation, on the one hand, and all the remaining summands, on the other, have different orders of smallness as $r \to 0$. Therefore, if we transform the rectangular coordinates x, y to the polar coordinates r, φ and then tend to the limit as $r \to 0$, we shall get

$$\left[1+(n-2)\frac{x^2}{r^2}\right]\Phi^{(2)}_{yy} - 2(n-2)\frac{xy}{r^2}\Phi^{(2)}_{xy} + \left[1+(n-2)\frac{y^2}{r^2}\right]\Phi^{(2)}_{xx} = 0.$$

This equation has however the only trivial solution $\Phi^{(2)} \equiv 0$ as follows from Theorem 8.3.

Assuming $\Phi^{(2)} \equiv 0$ in (9.6), we can rewrite the equation in the previous form, but the role of $\Phi^{(2)}$ will be played $\Phi^{(3)}$ and the summands under the summation sign will start from $k = 4$. To repeat the same reasoning, just as before, we get $\Phi^{(3)} \equiv 0$, and then $\Phi^{(k)} \equiv 0, k = 4, \ldots, p-1$.

Theorem 9.1 is proved.

10 Representation for $\delta L, \delta M, \delta N$ variations through component $\zeta(x,y)$ of a bending field

We suppose that $\tau = (\xi, \eta, \zeta)$ is a field of infinitesimal bendings of S, given by the equation $z = z(x,y)$. In the present section we establish the formulas for the variations $\delta L, \delta M, \delta N$ of the coefficients of the second quadratic form of the surface:

$$\delta L = \zeta_{xx}\sqrt{a}, \qquad \delta M = \zeta_{xy}\sqrt{a}, \qquad \delta N = \zeta_{yy}\sqrt{a}, \tag{10.1}$$

where a is the discriminant of the first quadratic form of S.

Denoting for convenience the coefficients L, M, N of the second quadratic form by $\pi_{11}, \pi_{12}, \pi_{22}$, we have [27]:

$$\pi_{ij} = \frac{1}{\sqrt{a}}(r_1, r_2, r_{ij}), \quad i,j = 1,2 \tag{10.2}$$

where $r = (x, y, z)$ is a radius vector of S, and differentiation with respect to x and y is denoted by indices 1 and 2.

According to the definition, $\tau = \delta r$. Therefore we get from (10.2)

$$\delta\pi_{ij} = \frac{1}{\sqrt{a}}\Big[(\tau_1, r_2, r_{ij}) + (r_1, \tau_2, r_{ij}) + (r_1, r_2, \tau_{ij})\Big]. \tag{10.3}$$

Let us extend the mixed product:

$$\begin{aligned}(\tau_1, r_2, r_{ij}) &= \xi_x z_{ij},\\ (r_1, \tau_2, r_{ij}) &= \zeta_y z_{ij},\\ (r_1, r_2, \tau_{ij}) &= \zeta_{ij} - z_x\xi_{ij} - z_y\eta_{ij}.\end{aligned}$$

After substituting this into (10.3) we have

$$\delta\pi_{ij} = \frac{1}{\sqrt{a}}\Big[\zeta_{ij} + (\xi_x + \eta_y)z_{ij} - z_x\xi_{ij} - z_y\eta_{ij}\Big]. \tag{10.4}$$

It remains to express $\xi_x, \eta_y, \xi_{ij}, \eta_{ij}$ through $\zeta(x, y)$, using (6.1):

$$\begin{aligned}&\zeta_x + \eta_y = -z_x\zeta_x - z_y\zeta_y,\\ &\xi_{xx} = -z_{xx}\zeta_x - z_x\zeta_{xx}, \qquad \eta_{xx} = -z_{xx}\zeta_y - z_y\zeta_{xx},\\ &\xi_{xy} = -z_{xy}\zeta_x - z_x\zeta_{xy}, \qquad \eta_{xy} = -z_{xy}\zeta_y - z_y\zeta_{xy},\\ &\xi_{yy} = -z_{yy}\zeta_x - z_x\zeta_{yy}, \qquad \eta_{yy} = -z_{yy}\zeta_y - z_y\zeta_{yy}.\end{aligned}$$

Transforming (10.4) by means of this relation we have (10.1).

Remark Since $z_{xx} = L\sqrt{a}, z_{xy} = M\sqrt{a}, z_{yy} = N\sqrt{a}$, then taking formulas (10.1) into account we can write equation (6.2) in the form

$$L\delta N - 2M\delta M + N\delta L = 0.$$

This relation is none other than the vanishing of the variation of Gaussian curvature under infinitesimal bendings of a surface.

11 Determination of deformation classes for the general surface in conjugate isometric coordinates

11.1 Let (x,y) and (u,v) be two different systems of coordinates on a surface. Let us denote the variations of coefficients of the second quadratic form in these coordinates by $\delta L(x,y), \delta M(x,y), \delta N(x,y)$ and $\delta L(u,v), \delta M(u,v), \delta N(u,v)$. Since these variations are the components of second valency tensors [27] then the connection between them is given by the formulas

$$\begin{aligned}\delta L(u,v) =& \left(\frac{\partial x}{\partial u}\right)^2 \delta L(x,y) + 2\frac{\partial x}{\partial u}\frac{\partial y}{\partial u}\delta M(x,y) + \left(\frac{\partial y}{\partial u}\right)^2 \delta N(x,y),\\ \delta M(u,v) =& \frac{\partial x}{\partial u}\frac{\partial x}{\partial v}\delta L(x,y) + \left(\frac{\partial x}{\partial u}\frac{\partial y}{\partial v} + \frac{\partial x}{\partial v}\frac{\partial y}{\partial u}\right)\delta M(x,y)\\ &+ \frac{\partial y}{\partial u}\frac{\partial y}{\partial v}\delta N(x,y).\\ \delta N(u,v) =& \left(\frac{\partial x}{\partial v}\right)^2 \delta L(x,y) + 2\frac{\partial x}{\partial v}\frac{\partial y}{\partial v}\delta M(x,y) + \left(\frac{\partial y}{\partial v}\right)^2 \delta N(x,y).\end{aligned} \tag{11.1}$$

If we change the places of the variable, i.e. $u \leftrightarrow x$ and $v \leftrightarrow y$, then we obtain the converse relations. All these relations help us to define a class of deformations in certain variables through the deformation class formulated in other variables.

Further we examine only surfaces of type S_1.

11.2 Let us apply formulas (11.1) to the model surface (6.3). We shall consider the conjugate isometric coordinates as (u,v) given by the formulas (2.2). Computing partial derivatives from (2.2) and introducing them into (11.1), we get

$$\begin{aligned}\delta L(u,v) =& \rho^{2\sqrt{n-1}-2}\left[\left\{1 + \left(\frac{1}{\sqrt{n-1}} - 1\right)\frac{u^2}{\rho^2}\right\}^2 \delta L(x,y)\right.\\ &+ 2\left\{1 + \left(\frac{1}{\sqrt{n-1}} - 1\right)\frac{u^2}{\rho^2}\right\}\left(\frac{1}{\sqrt{n-1}} - 1\right)\frac{uv}{\rho^2}\delta M(x,y)\\ &\left.+ \left(\frac{1}{\sqrt{n-1}} - 1\right)^2 \frac{u^2 v^2}{\rho^4}\delta N(x,y)\right].\end{aligned} \tag{11.2}$$

Here $\rho^2 = u^2 + v^2$. We shall not write the other two formulas as they are similar to (11.2) and give analogous properties for $\delta M(u,v)$ and $\delta N(u,v)$. Now if we use equalities (10.1), in which $a(x,y) = 1 + n^2 r^{2n-4}$, then from (11.2) we have

$$\delta L(u,v) = \rho^{2/\sqrt{n-1}-2}\sqrt{1 + n^2\rho^{2/\sqrt{n-1}}}\left[\left(1 + c\frac{u^2}{\rho^2}\right)^2 \zeta_{xx}\right.$$

$$+2\left(1+c\frac{u^2}{\rho^2}\right)\cdot c\frac{uv}{\rho^2}\zeta_{xy}+c^2\frac{u^2v^2}{\rho^4}\zeta_{yy}\Bigg]. \tag{11.3}$$

In this formula we replace r by ρ using the equality $\rho = r^{\sqrt{n-1}}$ and we use the notation $c = 1/\sqrt{n-1} - 1$.

Let us assume that $\zeta(x,y) \in C^p, p \geq 3$. On the basis of Bernstein's theorem [7] a bending field $\zeta(x,y)$ of surface (6.3) belongs to class C^A outside the flat point. It is clear that $\delta L(u,v), \delta M(u,v), \delta N(u,v)$ in the variables (u,v) belong to the same class outside $(0,0)$. As to the requirements of these variations at the flat point they follow from (11.2) with regard to Theorem 8.1:

$$\delta L(u,v), \delta M(u,v), \delta N(u,v) = O(\rho^{\nu_{k_{n,p}}+1/\sqrt{n-1}-2}), \quad \rho \to 0. \tag{11.4}$$

11.3 Now as in the previous subsection we clarify the conditions imposed on $\delta L(u,v), \delta M(u,v), \delta N(u,v)$ for surface (9.1). In the context of (u,v) we shall apply the conjugate isometric coordinates constructed in §3.

Let us suppose that $f(x,y) \in C^m(D)$ and $m > \max(3,n)$. Moreover let $\zeta(x,y) \in C^p(D), 3 \leq p \leq m$. Then thanks to Pogorelov's theorem [44], a bending field of $\zeta(x,y)$ of surface (9.1) belongs to class C_α^{m-1} outside the flat point (α is a Holder exponent arbitrarily close to 1). From (10.1) it follows that $\delta L(x,y), \delta M(x,y), \delta N(x,y) \in C_\alpha^{m-3}$. If we use the remark to Lemma 3.2, then from (11.1) we obtain

$$\delta L(u,v), \delta M(u,v), \delta N(u,v) \in C_\alpha^{m-3}$$

outside (0,0). It remains to learn the properties of these variations at the flat point. From (11.1) taking into account

$$\delta L(x,y), \delta M(x,y), \delta N(x,y) = O(r^{p-2}), \quad r \to 0,$$

and using Lemma 3.2 we have

$$\delta L(u,v), \delta M(u,v), \delta N(u,v) = O(\rho^{p/\sqrt{n-1}-2}), \quad \rho \to 0. \tag{11.5}$$

11.4 For the description of infinitesimal bendings of surface (9.1) in §4 we have introduced a complex–valued function by means of formula (4.15)

$$w(z) = z^2 K^{1/4}(z)(\delta M(z) + \mathrm{i}\delta L(z)). \tag{11.6}$$

To pass over to the habitual notation of complex analysis, here we denote again the conjugate isometric coordinates by $(x,y), z = x + iy$.

As $K(z) \in C_\alpha^{m-3}$ for $|z| > 0$ and $K(z) = O(|z|^{2(n-2)/\sqrt{n-1}})$, $z \to 0$, see (4.7), on the basis of the previous subsection we get $w(z) \in C_\alpha^{m-3}$ outside $z = 0$ and

$$w(z) = O(|z|^{(n-2+2p)/2\sqrt{n-2}}), \quad z \to 0. \tag{11.7}$$

11.5 Let us take some surface (9.1), belonging to the set of surfaces of type S_1. We suppose that $f(x,y) \in C^\infty$. If one considers its infinitesimal bendings from the regularity class C^∞, then $w(z)$ must have at $z = 0$ a zero of infinite order on account of the previous subsection. But then $w(z) \equiv 0$ on the whole surface; see Consequence 2.1 from §2 ch. 3.

Thus, the statement of Theorem 8.3 is spread to a wider class of surfaces:

Theorem 11.1 *Surfaces of type $S_1^{(\infty)}$ are locally rigid in the class C^∞.*

11.6 We denoted the function $w(z)$ for the model equation (6.3) by $\Phi(z)$. Outside the flat point, this function evidently belongs to class C^A. Its behaviour as $z \to 0$ is characterized by the formula

$$\Phi(z) = O(|z|^{(n-2+2\nu_{k_{n,p}+1})/2\sqrt{n-1})}, \quad z \to 0.$$

The asymptotic expression for $\Phi(z)$ differs from (11.7) because for surface (6.3) we use the more precise relation (11.4) in comparison to (11.5). If one remembers the notation for $\nu_{k_{n,p}+1}$ (see Theorem 8.1), we get

$$\Phi(z) = O(|z|^{\sqrt{((n-2)/2\sqrt{n-1})^2+(k_{n,p}+1)^2}}), \quad z \to 0. \tag{11.8}$$

12 General results for the model surface

Let us proceed to investigate the analytic problem equivalent to searching for infinitesimal bendings of the model surface with a flat point and a linear combination of the variations of the normal curvature and the geodesic torsion, given along a surface boundary. This problem consists of searching for solutions of the model equation

$$\partial_{\bar{z}}\Phi - \frac{\lambda}{2\bar{z}}\bar{\Phi} = 0, \quad |z| < R, \tag{12.1}$$

$$|\lambda| = \frac{n-2}{2\sqrt{n-1}}, \quad n > 2$$

satisfying the boundary condition

$$\mathrm{Re}(z^{-m}\Phi) = h(z), \quad |z| = R. \tag{12.2}$$

As we assume that a boundary field of the model surface belongs to class $C^p, p \geq 3$, then in addition to this, $\Phi(z)$ must be subject to relation (11.8):

$$\Phi(z) = O(|z|^{\mu_{k_{n,p}+1}}), \quad z \to 0. \tag{12.3}$$

where

$$\mu_{k_{n,p}+1} = \sqrt{|\lambda|^2 + (k_{n,p}+1)^2}$$

and $k_{n,p}$ is defined by (8.1).

The problem under consideration differs from the Riemann–Hilbert problem (see §9, ch. 3) by the supplementary requirement (12.3). To complete its solution it remains to obtain the conditions in which $\Phi(z)$ will possess a zero of order not less than $\mu_{k_{n,p}+1}$. Just as before we shall distinguish the cases $m = 0, m > 0$ and $m < 0$.

Theorem 12.1 $(m = 0)$ *The homogeneous boundary problem* (12.1) – (12.3) $(h(z) \equiv 0)$ *has only the zero solution. For the existence of a solution of the inhomogeneous problem it is necessary and sufficient that* $h(z)$ *should satisfy* $q_{n,p} = 2k_{n,p}+1$ *real solvability conditions*

$$\int_0^{2\pi} h(re^{i\varphi})\,d\varphi = 0, \qquad \int_0^{2\pi} h(re^{i\varphi}) \begin{pmatrix} \cos & k\varphi \\ \sin & k\varphi \end{pmatrix} d\varphi = 0, \quad (k = 1, \ldots, k_{n,p}). \tag{12.4}$$

This theorem is an obvious consequence of Theorem 9.1 from §9, ch. 2. As for conditions (12.4) they are extracted from the explicit form of solutions by ignoring the first summands whose vanishing orders are lower than the required value $\mu_{k_{n,p}+1}$.

Theorem 12.1 $(m > 0)$ *Let* $m > k_{n,p}$, *where* $k_{n,p}$ *is defined by equality (8.1). Then the homogeneouss boundary problem (12.1)– (12.3) has* $2(m - k_{n,p}) - 1$ *linearly independent solutions and the inhomogeneous problem is always solvable.*

Let $m \leq k_{n,p}$. *Then the homogeneous problem has only the zero solution, and for the existence of a solution of the inhomogeneous problem it is necessary and sufficient that* $h(z)$ *should satisfy* $2(k_{n,p} - m) + 1$ *real solvability conditions (12.4).*

Solving these inequalities with respect to $k_{n,p}$, we get (see (8.1)):

$$\sqrt{p(p+n-2)/(n-1)}-1<k_{n,p}\leq\sqrt{p(p+n-2)/(n-1)}. \tag{13.6}$$

Thus, if one real condition and $k_{n,p}$ complex conditions (1.7), ch. 3, are fulfilled then

$$w(z)=O(|z|^{\mu_{k_{n,p}+1}}),\quad z\to 0, \tag{13.7}$$

and therefore (13.3) is satisfied.

13.2 The scheme, indicated in subsection 13.1 and based on Theorem 1.1, ch. 3, is essentially used for solving problem (13.1)–(13.3) in every possible case (depending on the values of m). In fact, we must recall one of the stages in the investigation of the boundary problem (13.1), (13.2); see §§4–6, ch. 3. It consisted of the fact that the function $\Phi(z)$ in integral equation (13.4) is represented by the sum of two summands either of which is a solution of the model equation:

$$\Phi(z)=\Phi_O(z)+P_G^*\bar{w}. \tag{13.8}$$

Further, the matter was reduced to finding $\Phi_O(z)$ through $h(z)$ (in some cases with a defined arbitrariness, in other cases with preliminary restrictions on $h(z)$).

Equality (13.8) is however equivalent to the representation of the coefficients $a_k (k=0,1,\dots)$ in formula (13.5) by the sum:

$$a_k=a_k^{(0)}(h)+a_k^{(1)}(w), \tag{13.9}$$

where $a_k^{(0)}(h)$ and $a_k^{(1)}(w)$ are linear functionals, whose values are defined by functions $h(z)$ and $w(z)$.

Introducing the right side of (13.9) instead of a_k into relation (1.7) from ch. 3, we get those conditions which provide the realizability of (13.3) in each concrete case. For $m\geq 0$ some of them are satisfied by the choice of arbitrary constants, imposing restrictions upon $h(z)$.

Theorem 13.1 $(m=0)$ *Let* $\|\hat{P}_G\|_C<1$ *cf. Theorem 4.1, ch. 3). The homogeneous boundary problem (13.1)- (13.3)* $(h(z)\equiv 0)$ *has only the zero solution. For the*

Proof Let us examine the homogeneous problem (12.1)–(12.3). If we omit condition (12.3) then its general solution is given by formula (9.16) from §9 ch.2:

$$\Phi^*(z)=\Phi_O(z)+\sum_{q=1}^{m-1}\Phi_{m-q}(z)+\Phi_m(z).$$

Thanks to relations (9.11), (9.13), (9.15) from §9, ch. 2 the asymptotic behaviour of Φ_O,Φ_{m-q},Φ_m is characterized by the equalities

$$\Phi_O(z)=O(|z|^{\mu_o}),\qquad \Phi_m(z)=O(|z|^{\mu_m}),$$

$$\Phi_{m-q}(z)=O(|z|^{m-q}),\quad z\to 0\quad (q=1,\dots,m-1).$$

If $m>k_{n,p}$, then $\Phi_O(z)$ and $\Phi_{m-q}(z), q=m-k_{n,p},\dots,m-1$, will not satisfy (12.3), and we must cast them away. In this case the general solution of the homogeneous problem will have the form

$$\Phi^*(z)=\sum_{q=1}^{m-k_{n,p}-1}\Phi_{m-q}(z)+\Phi_m(z)$$

or (if one uses the notation of the general solution in the form (9.17) from §9, ch. 2):

$$\Phi^*(z)=\sum_{q=1}^{m-k_{n,p}-1}C_{m-q}^{(1)}\bar{\Phi}_{m-q}(z)+C_{m-q}^{(2)}\bar{\bar{\Phi}}_{m-q}(z)+C_m\bar{\Phi}_m(z),$$

where $C_{m-q}^{(1)},C_{m-q}^{(2)},C_m$ are arbitrary real constants.

If $m\leq k_{n,p}$, then none of the above solutions will satisfy condition (12.3) and therefore the homogeneous problem will have only the trivial solution.

Now we examine the inhomogeneous problem. In the case $m>k_{n,p}$ one of its solutions is evidently given by formula (9.18) from §9, ch.2:

$$H(z)=\sum_{q=O}^{\infty}H_q(z).$$

Since $H_q(z)=O(|z|^{\mu_{m+q}}), z\to 0$, then in the case $m\leq k_{n,p}$ to meet the condition (12.3) we should suppose $H_q(z)=0$ for $q=0,1,\dots k_{n,p}-m$. But this can be reached at the expense of $h_O=h_q=0(q=1,\dots,k_{n,p}-m)$, which is in turn equivalent to the solvability conditions (12.4). Theorem 12.1 $(m>0)$ is proved.

We should remark that Theorem 12.1 ($m > 0$) comprises the assertions of Theorem 12.1 ($m = 0$), and only for the convenience of the presentation were they formulated individually.

Theorem 12.1 ($m < 0$) *The homogeneous problem (12.1)–(12.3) has only the zero solution and to solve the inhomogeneous problem it is necessary and sufficient that $h(z)$ should satisfy $2(|m| + k_{n,p}) + 1$ real conditions.*

In fact, the statement, that the homogeneous problem has only the trivial solution, follows from Theorem 9.3 of §9, ch.2. As to the inhomogeneous problem, $h(z)$ must be subordinated to the $2|m|$ real conditions (9.19), ch. 2, with regard to the same theorem. After this, a solution of the inhomogeneous problem is given in the series form (19.20), ch. 2. Cutting the first $k_{n,p} + 1$ summands, whose zero orders are not less than those required in (12.3) at $z = 0$, off from this series we additionally get $2k_{n,p} + 1$ real conditions on $h(z)$.

Remark The supplementary solvability conditions, mentioned in the proof of Theorem 9.3 of ch. 2, have the form:

$$\mathrm{Re}\frac{1}{\mathcal{X}}\int_O^{2\pi} h(Re^{i\varphi})\Big[\sum_{\gamma=0}^{\infty}\frac{(-1)^\gamma}{\lambda^\gamma}e^{-i(2\gamma+1)|m|\varphi}\prod_{\delta=1}^{\gamma}P_{-2\delta|m|}\Big]\,d\varphi = 0,$$

$$\int_O^{2\pi} h(Re^{i\varphi})\Big[\sum_{\gamma=0}^{\infty}\frac{(-1)^\gamma}{\lambda^\gamma}e^{-i[2(\gamma+k+1)|m|\pm q]\varphi}$$

$$\times\prod_{\delta=1}^{\gamma}P_{-[(2\delta+2k+1)|m|\pm q]}\Big]\,d\varphi = 0,$$

where the integers $k, |m|, q$ satisfy the inequality $(2k+1)|m| \pm q \le k_{n,p}$; moreover, if one takes the plus sign, then q can take values $1, 2, \ldots, |m|$; if one takes the minus sign, then $q = 0, 1, \ldots, |m| - 1$. In addition, we assume (also in future) that $\prod_{\delta=1}^{O} = 1$.

13 Study of an analytic problem for the general surface

We proceed to the solution of the problem

$$\partial_{\bar z} w - \frac{b(z)}{2\bar z}\bar w = 0, \quad |z| < R, \tag{13.1}$$

$$\mathrm{Re}(z^{-m}w) = h(z), \quad |z| = R, \tag{13.2}$$

under the condition

$$|b(0)| = \frac{n-2}{2\sqrt{n-1}}, \quad n > 2.$$

We suppose that the surface and its bending field belong to the same regularity classes which were indicated for them in §11. In such a case $w(z)$ must satisfy condition (11.7):

$$w(z) = O(|z|^{(n-2+2p)/2\sqrt{n-1}}), \quad z \to 0. \tag{13.3}$$

The problem (13.1)–(13.3) is quite similar to the one which has been analysed in the previous section for the model surface. Its solution is hindered by technical obstacles. So as not to confuse matters, in the first subsection of this section we dwell on the mechanism of the solution to the problem, and in the following subsections we formulate the ultimate results.

We use the notation $\lambda = b(0)$.

13.1 Let $w(z)$ be some known solution of problem (13.1), (13.2). We must clarify the conditions which allow us to realize condition (13.3) on $w(z)$.

Being a solution of equation (13.1), $w(z)$, on account of Theorem 4.1 of ch. 1, must satisfy the integral equation (we assume $F(z) \equiv 0$) :

$$w(z) = \Phi(z) + P_G\bar w, \tag{13.4}$$

where $\Phi(z)$ is a solution of the model equation (12.1), represented by the series (see (9.4) from §9 ch. 2):

$$\Phi(z) = \mathcal{X}\cdot a_O\Big(\frac{r}{R}\Big)^{|\lambda|} + \sum_{k=1}^{\infty}(\lambda a_k e^{ik\varphi} + P_{-k}\bar a_k e^{-ik\varphi})\Big(\frac{r}{R}\Big)^{\mu_k}. \tag{13.5}$$

Since we suppose that $w(z)$ is given then $\Phi(z)$ is defined from (13.4) and the coefficients $a_O, a_k, k = 1, 2, \ldots$, becomes known from (13.5).

Now to obtain the relations guaranteeing conditions (13.3) it is necessary to use Theorem 1.1 of ch. 3. In fact, if we assume that equality (1.7) from ch. 3 holds, then

$$w(z) = O(|z|^{\mu_k}), \quad z \to 0,$$

where $\mu_k^2 = |\lambda|^2 + k^2, |\lambda| = (n-2)/2\sqrt{n-1}$. The choice of a proper number of relations of type (1.7), ch. 3, which are brought to the condition (13.3), is obviously realized from the inequalities

$$\mu_{k_{n,p}} \le (n-2+2p)/2\sqrt{n-1} < \mu_{k_{n,p}+1}.$$

existence of the inhomogeneous solution it is necessary and sufficient that $h(z)$ *should satisfy* $q_{n,p} = 2k_{n,p} + 1$ *real conditions:*

$$\int_O^{2\pi} h(Re^{i\varphi})\,d\varphi = \frac{Re\mathcal{X}}{\mathcal{X}} \iint_G \left[\frac{b(\zeta) - b(0)}{2|\zeta|^2}\overline{w(\zeta)} + \frac{\lambda}{|\lambda|}\frac{\overline{b(\zeta)} - \overline{b(0)}}{2|\zeta|^2}w(\zeta)\right]\left(\frac{R}{|\zeta|}\right)^{|\lambda|} d\xi\,d\eta$$
$$+ \iint_G \left[\frac{|\lambda| - \bar{\lambda}}{4|\lambda|}\cdot\frac{b(\zeta) - b(0)}{2|\zeta|^2}\overline{w(\zeta)} + \frac{|\lambda| - \lambda}{4|\lambda|}\frac{\overline{b(\zeta)} - \overline{b(0)}}{|\zeta|^2}w(\zeta)\right]\left(\frac{|\zeta|}{R}\right)^{|\lambda|} d\xi\,d\eta,$$

$$\int_O^{2\pi} h(Re^{i\varphi})e^{-ik\varphi}d\varphi = \frac{\lambda + P_{-k}}{4\lambda}\iint_G \left[\frac{P_k}{\mu_k}\frac{b(\zeta) - b(0)}{|\zeta|^2}\overline{w(\zeta)} + \frac{\lambda}{\mu_k}\frac{\overline{b(\zeta)} - \overline{b(0)}}{|\zeta|^2}w(\zeta)\right]\left(\frac{R}{|\zeta|}\right)^{\mu_k} e^{-ik\gamma}d\xi\,d\eta$$
$$+ \iint_G \left[\frac{P_{-k} - \lambda}{4\mu_k}\frac{b(\zeta) - b(0)}{|\zeta|^2}\overline{w(\zeta)} + \frac{P_k - \lambda}{4\mu_k}\frac{\overline{b(\zeta)} - \overline{b(0)}}{|\zeta|^2}w(\zeta)\right]\left(\frac{|\zeta|}{R}\right)^{\mu_k} e^{-ik\gamma}\,d\xi d\eta,$$

where $k = 1, 2, \ldots, k_{n,p}$ *and it is necessary to use the relation (4.6) of ch. 3 for* $\lambda = b(0) < 0$ *instead of the first real condition. All the notation, applied here, is borrowed from ch. 3.*

The case $m > 0$ *Let* $\|P_G^*\|_C < 1$ *(cf. Theorem 5.1, ch.3).*

If $m > k_{n,p}$*, where* $k_{n,p}$ *is defined by (13.6), then the homogeneous problem (13.1)-(13.3) has* $2(m - k_{n,p}) - 1$ *linearly independent real solutions, and the inhomogeneous problem is always solvable.*

If $m \geq k_{n,p}$*, then the homogeneous problem has only the trivial solution, and for the existence of a solution of the inhomogeneous problem it is necessary and sufficient that* $h(z)$ *should satisfy* $2(k_{n,p} - m) + 1$ *real conditions:*

$$R^m \int_O^{2\pi} h(Re^{i\varphi})\,d\varphi = Re \iint_G \left[\frac{P_m}{2\mu_m} \frac{b(\zeta) - b(0)}{|\zeta|^2} \overline{w(\zeta)} \right.$$
$$\left. + \frac{\lambda}{2\mu_m} \frac{\overline{b(\zeta)} - \overline{b(0)}}{2|\zeta|^2} w(\zeta) \right] \cdot \left(\frac{R}{\rho} \right)^{\mu_m} e^{-im\varphi} d\xi\, d\eta$$
$$+ \iint_G \left[\alpha_O(\zeta) \frac{b(\zeta) - b(0)}{|\zeta|^2} \overline{w(\zeta)} + \beta_O(\zeta) \frac{\overline{b(\zeta)} - \overline{b(0)}}{|\zeta|^2} w(\zeta) \right] d\xi\, d\eta,$$

$$R^m \sum_{l=O}^{[k/2m]} \frac{1}{(-\lambda)^l} \prod_{\delta=1}^{l} P_{-[(2\delta-1)m+k-2ml]} \int_O^{2\pi} h(Re^{i\varphi}) e^{-i(k-2ml)\varphi}\, d\varphi$$
$$= \iint_G \left[\frac{P_{m+k}}{4\mu_{m+k}} \frac{b(\zeta) - b(0)}{|\zeta|^2} \overline{w(\zeta)} \right.$$
$$\left. + \frac{\lambda}{4\mu_{m+k}} \frac{\overline{b(\zeta)} - \overline{b(0)}}{|\zeta|^2} w(\zeta) \right] \cdot \left(\frac{R}{|\zeta|} \right)^{\mu_{m+k}} e^{-i(m+k)\varphi}\, d\xi\, d\eta,$$
$$+ \iint_G \left[a_k(\zeta) \frac{b(\zeta) - b(0)}{2|\zeta|^2} \overline{w(\zeta)} + \beta_k(\zeta) \frac{\overline{b(\zeta)} - \overline{b(0)}}{2|\zeta|^2} w(\zeta) \right] d\xi\, d\eta$$

where $k = 1, 2, \ldots, k_{n,p} - m$ and

$$a_O(\zeta) = \frac{1}{2}[N_m(\zeta) + \overline{M_m(\zeta)}],$$
$$a_k(\zeta) = N_{m+k}(\zeta) + \overline{M_{m-k}(\zeta)}$$
$$+ \sum_{l=1}^{[k/2m]} (N_{m+k-2ml} + \bar{M}_{m-k+2ml}) \frac{(-1)^l}{\lambda^l} \prod_{\delta=1}^{l} P_{-[(2\delta-1)m-k-2ml]}.$$

In addition, $[k/2m]$ *denotes the integer part of* $k/2m$ *and the last sum is equal to zero for* $k < 2m$. *The formulas for* $\beta_k(\zeta)$ *are obtained from the formula for* $a_k(\zeta)$ *by replacing the letter* N *by* M, *and conversely (all other notation is borrowed from §5, ch. 3).*

The case $m < 0$. *The homogeneous problem (13.1) – (13.3) has only the trivial solution. For solvability of the inhomogeneous problem with* $\|P_G^{**}\|_C < 1$ *(see Theorem 6.1, ch. 3) it is necessary and sufficient that* $h(z)$ *should satisfy* $2(|m| + k_{n,p}) + 1$

real conditions, from which $2|m|$ *conditions are given by equalities (6.3) of §6, ch. 3, and the rest have the form:*

$$\mathcal{X} R^m \operatorname{Re}\left[\frac{1}{\mathcal{X}} \int_0^{2\pi} h(Re^{i\varphi}) \left\{ \sum_{\beta=0}^{\infty} \frac{(-1)^\beta}{\lambda^\beta} e^{-i(2\beta+1)|m|\varphi} \prod_{\delta=1}^{\beta} P_{-2\delta|m|} \right\} d\varphi \right]$$

$$= \iint_G \left[\left(\frac{\bar{\mathcal{X}} B_{O,|m|}(\zeta) + \mathcal{X} \overline{C_{O,|m|}(\zeta)}}{\bar{\mathcal{X}} R^{|\lambda|}} + \left(\frac{R}{|\zeta|}\right)^{|\lambda|} \right) \frac{\beta(\zeta) - b(0)}{4|\zeta|^2} \overline{w(\zeta)} \right.$$

$$+ \left(\frac{\bar{\mathcal{X}} C_{O,-|m|}(\zeta) + \mathcal{X} \overline{B_{O,-|m|}(\zeta)}}{\bar{\mathcal{X}} R^{|\lambda|}} \right.$$

$$\left. \left. + \frac{\lambda}{|\lambda|} \left(\frac{R}{|\zeta|}\right)^{|\lambda|} \right) \frac{\overline{b(\zeta)} - \overline{b(0)}}{4|\zeta|^2} w(\zeta) \right] d\xi \, d\eta,$$

$$R^m \int_0^{2\pi} h(Re^{i\varphi}) \left[\sum_{\beta=0}^{\infty} \frac{(-1)^\beta}{\lambda^\beta} e^{-i[2(\beta+k+1)|m|\pm q]\varphi} \prod_{\delta=1}^{\beta} P_{-[2\beta+k+1)|m|\pm q]} \right] d\varphi$$

$$= \iint_G \left\{ \frac{P_{(2k+1)|m|\pm q}}{4\mu_{(2k+1)|m|\pm q}} \frac{b(\zeta) - b(0)}{|\zeta|^2} \overline{w(\zeta)} + \frac{\lambda}{4\mu_{(2k+1)|m|\pm q}} \right.$$

$$\times \left. \frac{\overline{b(\zeta)} - \overline{b(0)}}{|\zeta|^2} w(\zeta) \right\} \cdot \left(\frac{R}{|\zeta|}\right)^{\mu_{(2k+1)|m|\pm q}} e^{-i[(2k+1)|m|\pm q]\gamma} \, d\xi \, d\eta$$

$$+ \iint_G \left[\left(\frac{\lambda P_{-[(2k+1)|m|\pm q]}}{2\mu_{(2k+1)|m|\pm q}} \overline{A_{k,\pm q}(\zeta)} + B_{k,\pm q}(\zeta) \right) \frac{b(\zeta) - b(0)}{2|\zeta|^2} \overline{w(\zeta)} \right.$$

$$+ \left(-\frac{\lambda^2}{2\mu_{(2k+1)|m|\pm q}} \overline{A_{k,\pm q}(\zeta)} + C_{k,\pm q}(\zeta) \right)$$

$$\times \left. \frac{\overline{b(\zeta)} - \overline{b(0)}}{2|\zeta|^2} w(\zeta) \right] R^{\mu_{(2k+1)|m|\pm q}} \, d\xi \, d\eta,$$

where the integers $k, |m|, q$ *satisfy the inequalities* $(2k+1)|m| \pm q \leq k_{n,p}$*; moreover if one takes the plus sign, then* q *can take values, 1, 2, ...,* $|m|$*, and if one minus sign, then* $q = 0, 1, \ldots, |m| - 1$ *the notation, utilized here, is borrowed from §6, ch. 3).*

Remark 1 It should be indicated that the solvability conditions of theorem 13.1 although containing the unknown function w, in fact impose restrictions only on $h(z)$. As a matter of fact, $w(z)$ as a solution of the corresponding integral equation, if uniquely expressed through $h(z)$.

Remark 2 The comparison of these results with those for problem (13.1), (13.2) (see Theorems 4.1, 5.1, 6.1 ch. 3) shows that condition (13.3) imposes $2k_{n,p}+1$ supplementary restrictions upon the variety of solutions.

14 The influence of a flat point on infinitesimal bendings of the surface with boundary conditions

We recall the principal assumptions, made by us in the solution of the corresponding analytic problem before proceeding to the analysis of the geometric case.

14.1 In rectangular Cartesian coordinates $Oxyz$ the surface S is given by the equation

$$z = (x^2+y^2)^{n/2} f(x,y), \quad n > 2. \tag{14.1}$$

It is assumed that the flat point $x = y = 0$ is interior for a simply connected domain D, bounded with a Ljapunov contour, and $f(x,y) \in C^q(D)$, where $f(0,0) > 0$ and q is a sufficiently large integer (in particular, $q = \infty$) and $q > \max(n,3)$. Moreover, the Gaussian curvature of the surface S is positive for all points of S different from (0,0).

The set of surfaces (14.1) is divided into two subsets of surfaces of types S_O and S_1 according to the construction of §7: the first subset is countable and the second one is a continuum. All the following results refer only to type S_1 surfaces.

Since the compactness conditions on the corresponding integral operators are essentially utilized in the analytic apparatus, then this fact imposes, in turn, certain restrictions on the initial surface S_1: either its size is in a sense small or it differs little from the model surface

$$z = (x^2+y^2)^{n/2}, \quad n > 2.$$

Further, it is supposed that a bending field of the surface S_1 belongs to the regularity class $C^p(D), 3 \le p < q$. In addition such infinitesimal bendings of S_1 can be found which satisfy a linear combination of the variation of the normal curvature δk_s and the geodesic torsion $\delta\tau_s$ given along the boundary curve:

$$\lambda(s)\delta k_s + \mu(s)\delta\tau_s = h(s), \tag{14.2}$$

where $\lambda(s), \mu(s), h(s)$ are given functions of the curve length and are Holder continuous. Condition (14.2) is also equivalent to a fully defined linear combination of the variations of the coefficients of the second quadratic forms $\delta L, \delta M, \delta N$. If we take relation (10.1) into account, then (14.2) also corresponds to a linear combination of the second-order derivatives from the vertical component $\zeta(x, y)$ of the bending field vector.

Let us introduce the complex-valued function $g = \lambda\bar{\gamma}_1 + \mu\bar{\gamma}_2(\gamma_1, \gamma_2$ are defined by formula (5.1)). We denote the increment in the argument of the function g with a single circuit of the boundary curve, divided by 2π, by n.

Now we formulate the summarizing theorem which gives the solution of the geometric problem. It is an evident consequence of Theorem 13.1.

14.2 Theorem 14.1 *Let $m > k_{n,p}$. Then any surface of type S_1 admits a $2(m - k_{n,p}) - 1$ parametric set of infinitesimal bendings satisfying boundary condition (14.2).*

Let $m \leq k_{n,p}$. The surface S_1 is rigid if its deformations along the boundary are subordinated to the homogeneous condition $(h(s) \equiv 0)$, and has a unique infinitesimal bending for $h(s) \neq 0$ if $h(s)$ satisfies $2(k_{n,p} - m) + 1$ real conditions.

Remark 1 As in condition (14.2) [5], let us consider particular cases of

$$1. \quad \delta k_s = 0; \qquad 2. \quad \delta\tau_s = 0; \qquad 3. \quad \delta k_s = \delta\tau_s.$$

The rigidity of a type S_1 surface under any of those relations follows from Theorem 14.1.

We should note that the rigidity phenomenon of a surface on which points of flattening may exist, with a stationary normal curvature along the boundary, was stated in a different way in [73].

Remark 2 It follows from the results of I. Vekua concerning the solution of the Riemann–Hilbert problem [69] that a unique infinitesimal bending of a surface of a strictly positive curvative $(K \geq k_O > 0)$ under the condition of $\delta k_s = h(s)$ (or $\delta\tau_s = h(s)$) on the boundary, exists if and only if $h(s)$ is subordinated to three real conditions. In this case the number of conditions is equal to $2k_{n,p} + 1$. Consequently, the difference is equal to $2(k_{n,p} - 1)$.

Remark 3 We should note that the assertion of Theorem 14.1 in this case of $h(s) = 0$ and $m < 0$ is valid without restrictions on the distances of S_1 (see Theorem 13.1, the case $m < 0$).

Remark 4 Let us analyse the number $k_{n,p}$ which figures in Theorem 14.1 and is defined by inequality (13.6).We take a concrete surface of type S_1 (n is fixed) and consider its deformation from different regularity classes. It follows from (13.6) and Theorem 14.1, that increasing the regularity of a bending field (i.e. $p \to \infty$) leads to increasing the number of restrictions ($k_{n,p} \to \infty$), imposed on the existence of infinitesimal bendings of type S_1 surfaces.

Thus, the available flat point strengthens the resistance of the surface in respect of infinitesimal bendings.

15 Surfaces with a more complicated structure in a neighbourhood of a flat point

Let S be a sufficiently regular surface with positive curvature everywhere except at some point O, which is a flat point. We introduce a Cartesian system of coordinates $Oxyz$ with the origin at the flat point and axes x, y situated in the tangent plane of S. Then the surface in a neighbourhood of the point O can be given by the equation

$$z = \sum_{k=O}^{n} a_{k,n-k} x^k y^{n-k} + R(x,y) \tag{15.1}$$

where $n \geq 3$ is an integer and

$$a_{k,n-k} = \frac{1}{n!} \frac{\partial^n z(0,0)}{\partial x^k \partial y^{n-k}},$$

The function $R(x,y)$ belongs to the same regularity class as S, and $R(x,y) = O[(x^2 + y^2)^{(n+1)/2}]$ as $x, y \to 0$. Passing over to polar coordinates ($x = r\cos\varphi, y = r\sin\varphi$), we write (15.1) in the form

$$z = r^n f(\varphi) + R(x,y), \tag{15.2}$$

where

$$f(\varphi) = \sum_{k=O}^{n} a_{k,n-k} (\cos\varphi)^k (\sin\varphi)^{n-k}.$$

The requirement of a positive curvature in a neighbourhood of the point O imposes on $f(\varphi)$ the restriction

$$\Delta(\varphi) = (n-1)[-(n-1)f'^2(\varphi) + nf(\varphi)f''(\varphi) + n^2 f^2(\varphi)] > 0. \tag{15.3}$$

In addition we can let $f(\varphi) > 0$.

The first summand on the right side of (15.2) defines the structure of the surface in a neighbourhood of the flat point. The model surface

$$z = r^n f(\varphi) \tag{15.4}$$

corresponds to it.

In future we shall consider the surface (15.2) and (15.4) with looser restrictions on n and $f(\varphi)$. We assume that $n > 2$ is any real number, and $f(\varphi)$ is a $2\pi-$periodic function from the class $C^3[0, 2\pi]$, satisfying inequality (15.3).

It is easy to see that the surfaces under study are objects of a more complicated structure compared to those which had been studied previously. The aim of this section is to obtain the equation of infinitesimal bendings for surfaces (15.2) and (15.4) by a familiar method.

15.1 Let us construct conjugate isometric coordinates on surface (15.4). This problem is equivalent to finding a homeomorphic solution of the Beltrami equation

$$\partial_{\bar{\zeta}} w - q(\zeta)\partial_{\zeta} w = 0, \quad \zeta \in D, \tag{15.5}$$

where $\zeta = x + \mathrm{i}y = r\mathrm{e}^{\mathrm{i}\varphi}, w = u + \mathrm{i}v, \quad \mathrm{i}^2 = -1,$

$$\partial_{\bar{\zeta}} = \frac{1}{2}\left(\frac{\partial}{\partial x} + \mathrm{i}\frac{\partial}{\partial y}\right), \qquad \partial_{\zeta} = \frac{1}{2}\left(\frac{\partial}{\partial x} - \mathrm{i}\frac{\partial}{\partial y}\right),$$

D is a certain neighbourhood of the point (0,0) in the complex plane ζ and

$$q(\zeta) = \frac{z_{xx} - z_{yy} + 2\mathrm{i}z_{xy}}{z_{xx} + z_{yy} + 2\sqrt{z_{xx}z_{yy} - z_{xy}^2}}.$$

Computing $q(\zeta)$ for surface (15.4), we get

$$q(\zeta) = q_O(\varphi)\mathrm{e}^{2\mathrm{i}\varphi}, \tag{15.6}$$

where

$$q_O(\varphi) = \frac{n(n-2)f(\varphi) - f''(\varphi) + 2\mathrm{i}(n-1)f'(\varphi)}{n^2 f(\varphi) + f''(\varphi) + 2\sqrt{\Delta(\varphi)}}.$$

From (15.6) we have $q(\zeta) \in C^1(D\backslash(0,0))$. Applying (15.3) we state

$$|q(\zeta)| = |q_O(\varphi)| < 1.$$

Let us assume

$$\mathcal{X} = 2\pi n(n-1) / \int_O^{2\pi} \frac{\sqrt{\Delta(\theta)}}{f(\theta)}\, \mathrm{d}\theta,$$

$$\gamma(\varphi) = \gamma_1(\varphi) + \mathrm{i}\gamma_2(\varphi),$$

where

$$\gamma_1(\varphi) = \frac{1}{n(n-1)} \int_O^{\varphi} \frac{\sqrt{\Delta(\theta)}}{f(\theta)}\, \mathrm{d}\theta > 0,$$

$$\gamma_2(\varphi) = -\frac{1}{n} \ln f(\varphi), \quad f(\varphi) > 0.$$

Lemma 15.1 *The function*

$$w(\zeta) = r^{\mathcal{X}} \mathrm{e}^{\mathrm{i}\mathcal{X}\gamma(\varphi)} \tag{15.7}$$

is a particular solution of the Beltrami equation (15.5), (15.6) realizing a homeomorphic mapping of domain D of the complex plane $\zeta = x + \mathrm{i}y$ onto some neighbourhood D_w of the point $w = 0$ of the plane $w = u + \mathrm{i}v, \zeta = 0$ corresponding $w = 0$, and conversely.

We do not prove this lemma because it does not contain any difficulties.

15.2 Now we construct conjugate infinitesimal coordinates on the surface (15.2). In this case the Beltrami equation takes the form

$$\partial_{\bar{\zeta}} U - q^*(\zeta)\partial_{\zeta} U = 0, \quad \zeta \in D. \tag{15.8}$$

Here $U = U_1 + \mathrm{i}U_2$ and $q^*(\zeta) = q(\zeta) + q_1(\zeta)$, where $q(\zeta)$ is defined by equality (15.6) and $q_1(\zeta)$ is a known function, possessing the properties $|q_1(\zeta)| < 1$ for all $\zeta \in D$ and $q_1(\zeta) = O(|\zeta|), \zeta \to 0$.

Replacing the variable ζ by w from formula (15.7) in equation (15.8), we get

$$\partial_{\bar{w}} U - G(w)\partial_w U = 0, \quad w \in D_w, \tag{15.9}$$

where

$$G(w) = \frac{q_1(\zeta)}{1 - |q(\zeta)|^2 - q_1(\zeta)\overline{\overline{q(\zeta)}}} \frac{w_\zeta}{\overline{(w_\zeta)}},$$

where $G(w) \in C^1$ outside $w = 0$ and $G(w) = O(|w|^{1/\mathcal{X}}), w \to 0$. From the last equation it follows that $|G(w)| \leq G_O < 1$ at least in some neighbourhood of $w = 0$. But then (15.9) admits a homeomorphic solution $U^*(w)$ in a neighbourhood of $w = 0$, see [69].

Analyzing the function $U^*(w(\zeta))$ where $w(\zeta)$ is given by equality (15.7), we arrive at the homeomorphic solution of equation (15.9) and therefore at the conjugate isometric coordinates on surface (15.2) in a neighbourhood of the flat point.

15.3 Now we proceed to the equation of infinitesimal bendings, see (1.2):

$$\partial_{\bar{w}} U + \overline{B(w)} \bar{U} = 0, \tag{15.10}$$

where

$$B(w) = -e^{i\psi(w)} \partial_{\bar{w}} (\operatorname{Arch} \frac{H}{\sqrt{K}}) - \frac{\partial_w K}{4K} e^{2i\psi(w)} \tag{15.11}$$

and

$$e^{i\psi(w)} = \frac{a^-(w)\sqrt{K(w)}}{2\sqrt{a(w)\mathcal{F}(w)}}$$

Let us compute for model surface (15.4) the singularity of $B(w)$ at the point $w = 0$, which corresponds to the flat point. As in §4, we calculate the necessary geometric quantities in the initial system of coordinates $\zeta = x + iy$:

$$\begin{aligned}
a^+(\zeta) &= 2 + r^{2n-2}(n^2 f^2 + f'^2), \\
a^-(\zeta) &= r^{2n-2}(n^2 f^2 + 2inff' - f'^2)e^{2i\varphi}, \\
a(\zeta) &= 1 + r^{2n-2}(n^2 f^2 + f'^2), \\
K(\zeta) &= \frac{r^{2n-4}}{a^2}\Delta, \\
H(\zeta) &= \frac{r^{n-2}}{2a^{3/2}}(n^2 f + f'' + r^{2n-2}\frac{n}{n-1} f\Delta), \\
&(\Delta = (n-1)[-(n-1)f'^2 + nff' + n^2 f^2]), \\
\frac{H(\zeta)}{\sqrt{K(\zeta)}} &= \frac{1}{2a^{1/2}} \cdot \frac{n^2 f + f'' + r^{2n-2}(n/(n-1))f\Delta}{\sqrt{\Delta}}, \\
\mathcal{F}(\zeta) &= H^2 - K = r^{2n-4}\left\{\frac{[n(n-2)f - f'']^2}{4} + (n-1)^2 f'^2 + r^{2n-2} g_1(r,\varphi)\right\}, \\
\frac{K(\zeta)}{\mathcal{F}(\zeta)} &= \frac{4\Delta}{[n(n-2)f - f'']^2 + 4(n-1)^2 f'^2} + r^{2n-2} g_2(r,\varphi), \\
&g_1(r,\varphi),\quad g_2(r,\varphi) = O(1), \quad r \to 0.
\end{aligned}$$

Using these expressions and the formulas of the transformation for the geometric quantities (see the details in §4) from the old coordinates $\zeta = x + iy$ to the new ones $w = u + iv$ according to equality (15.7), we get

$$\begin{aligned}
e^{i\psi(w)} &= -\frac{1+\sqrt{\Delta}/n(n-1)f - if'/nf}{1+\sqrt{\Delta}/n(n-1)f + if'/nf} \\
&\times \frac{n(n-2)f - f'' + 2i(n-1)f'}{\sqrt{[n(n-2)f - f'']^2 + 4(n-1)f'^2}} \cdot e^{2i\mathcal{X}\gamma_1} + O(r^{2n-2}), \\
\left(\mathrm{Arch}\frac{H}{\sqrt{K}}\right) &= \frac{in(n-1)f(\varphi)}{2\mathcal{X}r^{\mathcal{X}}} e^{ik\bar{\gamma}} \\
&\times \left[\frac{((n^2 f + f''/\sqrt{\Delta})_\varphi}{\sqrt{[n(n-2)f - f'']^2 + 4(n-1)^2 f'^2}} + O(r^{2n-2})\right], \\
\frac{1}{4}\partial_w \ln K &= \frac{|n(n-1)f + \sqrt{\Delta} + i(n-1)f'|^2}{4\sqrt{\Delta}(n^2 f + f'' + 2\sqrt{\Delta})} \frac{r^{-\mathcal{X}}}{2\mathcal{X}} e^{-i\mathcal{X}\gamma} \\
&\times \left[2(n-2)\frac{\sqrt{\Delta} + i(n-1)f'}{n(n-1)f} - i\frac{\Delta_\varphi}{\Delta} + O(r^{2n-2})\right], \quad r \to 0.
\end{aligned}$$

Substituting these expressions into formulas (15.11) and carrying out the appropriate transformations, we obtain

$$B(w) = \frac{1}{2\mathcal{X}}\left(\frac{w}{\bar{w}}\right)^2 \frac{\overline{b(w)}}{w},$$

where

$$\overline{b(w)} = \beta(\varphi) + b^*(w), \tag{15.12}$$

and $b^*(w) = O(|w|^{2(n-1)/\mathcal{X}}), w \to 0$, and

$$\begin{aligned}
\beta(\varphi) =& \frac{n(n-1)f + \sqrt{\Delta} - i(n-1)f'}{n(n-1)f + \sqrt{\Delta} + i(n-1)f'} \quad \frac{n(n-1)f}{n(n-1)f - f'' - 2i(n-1)f'} \\
&\times \left\{ i\left(\frac{n^2 f + f''}{\sqrt{\Delta}}\right)_\varphi - \left[2(n-2)\frac{\sqrt{\Delta} + i(n-1)f'}{n(n-1)f} - i\frac{\Delta_\varphi}{\Delta}\right] \right. \\
&\times \left. \frac{n(n-1)f + \sqrt{\Delta} - i(n-2)f'}{n(n-1)f + \sqrt{\Delta} + i(n-1)f'} \cdot \frac{n(n-2)f - f'' + 2i(n-1)f''}{4\sqrt{\Delta}} + O(r^{2n-2}) \right\}, \\
&r \to 0.
\end{aligned}$$

We should note that the function $\beta(\varphi)$, $\varphi = \arg \zeta$, must be expressed through the variable $\psi = \arg w$. The necessary condition is extracted from (15.7) and has the form

$$\psi = \mathcal{X}\gamma_1(\varphi) \equiv \frac{\mathcal{X}}{n(n-1)} \int_O^{\varphi} \frac{\sqrt{\Delta(\theta)}}{f(\theta)} \mathrm{d}\,\theta$$

Introducing $B(w)$ into (15.10), we have

$$\partial_{\bar{w}} U - \frac{1}{2\mathcal{X}} \left(\frac{\bar{w}}{w}\right)^2 \frac{b(w)}{\bar{w}} \bar{U} = 0.$$

If we pass over to the new unknown function

$$U^*(w) = w^2 U(w) = w^2 K^{1/4}(w)(\delta M + \mathrm{i}\delta L),$$

then the equation takes the final form

$$\partial_{\bar{w}} U^* - \frac{b(w)}{2\mathcal{X}\bar{w}} \bar{U}^* = 0,$$

and according to (15.12) we have

$$b(0) = \overline{\beta(\varphi)}.$$

15.4 Let us summarize the above, applying the customary notation. The infinitesimal bendings of surface (15.4) and (15.2) are also described by the complex–valued function

$$w(z) = z^2 K^{1/4}(z)(\delta M(z) + \mathrm{i}\delta L(z)),$$

which satisfies the equation

$$\partial_{\bar{z}} w - \frac{\lambda(\varphi) + b_O(z)}{2\bar{z}} \bar{w} = 0, \quad z \in G, \tag{15.13}$$

where the point $z = 0$ is within the domain $G, \lambda(\varphi)$ is a $2\pi-$periodic continuous function, $b_O(z)$ is continuous in G; moreover $b_O(z) = O(|z|^\alpha), \alpha > 0, z \to 0$ (for more detailed information on $\lambda(\varphi)$ see subsection 15.3).

The equation

$$\partial_{\bar{z}} \Phi - \frac{\lambda(\varphi)}{2\bar{z}} \bar{\Phi} = 0, \quad z \in G, \tag{15.14}$$

is the model equation in relation to (15.13). For $\lambda(\varphi) = \lambda_O$ and $\lambda(\varphi) = \lambda_O e^{im\varphi}(\lambda_O)$ is a complex number and m is an integer) equations (15.3), (15.4) are reduced to the particular cases studied in ch. 1–4.

16 Additions to Chapter 6

The surfaces (3.1) and (15.2), examined in the present chapter, had a contact of finite order with their tangent plane at the flat point. The infinitesimal bendings of such surfaces, referred to conjugate isometric coordinates, had been described by generalized Cauchy–Riemann systems, their coefficients having first order poles at a point corresponding to the isolated flat point of the surfaces.

In this connection there is a natural interest in the equations of infinitesimal bendings for such surfaces that have an infinite order of contact with its tangent plane at a flat point. One such object will be examined in the present section [2].

Let the radius vector of a surface of revolution be given in the form

$$r(\xi, \theta) = \varphi(\xi)\tau(\theta) + \xi k, a \leq \xi \leq b, \quad 0 \leq \theta \leq 2\pi$$

where the mutually perpendicular unit vectors τ and k (the latter being directed along the axis of revolution define the plane of the meridian; θ is the angle between this plane and some fixed one passing through the axis of revolution and $\varphi(\xi)$ is the equation of the meridian.

We suppose that $\varphi(\xi) \in C[a, b], \varphi(a) = 0$ and $\varphi(\xi) \in C^3, \varphi(\xi) > 0, \varphi''(\xi) = d^2\varphi/d\xi^2 < 0$ for $a < \xi \leq b$. We shall examine the function $\xi = \omega(\varphi)$, the inverse of $\varphi = \varphi(\xi)$, in neighbourhood of the point $\xi = a$. We assume that $\xi = \omega(\varphi)$ has the representation

$$\omega(\varphi) = \omega(0) + e^{-\alpha\varphi^{-4\beta}}\omega_O(\varphi), \tag{16.1}$$

where α and β are arbitrary positive numbers, $\omega_O(\varphi) \in C^3$ in a closed neighbourhood of the point $\varphi = 0$, and $\omega_O(0) \neq 0$.

In this way we deal with a simply–connected portion of a surface of revolution, with a pole at the point $\xi = a(\varphi = 0)$, and with an edge consisting of a parallel at $\xi = b$. Such a surface has a positive curvature everywhere except at the pole, where there is an isolated point of flattening. In view of (16.1) the order of contact with the tangent plane at the pole is infinite.

Now we construct conjugate isometric coordinates $z = x + \mathrm{i}y = r\exp(\mathrm{i}\theta)$ on this surface and as in [3] we relate the old coordinates (ξ, θ) and the new coordinates (x, y) by the equality

$$z = \exp\Big(- \int_\xi^b \sqrt{-\frac{\varphi''(\xi)}{\varphi(\xi)}}\,\mathrm{d}\xi + \mathrm{i}\theta \Big). \tag{16.2}$$

It is clear that the homeomorphism (16.2) maps the surface onto the interior of a unit circle in the z plane with centre at the origin. The edge of the surface is mapped onto the circumference $|z| = 1$, and the pole goes into the point $z = 0$. To show the latter, we go over, in some neighbourhood of the pole $\varphi = 0$, from $\varphi = \varphi(\xi)$ to the inverse function $\xi = \omega(\varphi)$ and use the representation (16.1). Then we obtain from (16.2)

$$r = |z| = \exp\Big(- \sqrt{-\frac{\alpha}{\beta}}\varphi^{-2\beta} + a_O + O(\varphi^{2\beta}) \Big), \quad \varphi \to 0, \tag{16.3}$$

where O is the 'big O' notation and a_O is a constant.

In conjugate isometric coordinates $z = x + \mathrm{i}y = r\mathrm{e}^{\mathrm{i}\theta}$, infinitesimal bendings of our surface are described by the equation (see [38]):

$$\partial_{\bar z} w + \frac{\mathrm{e}^{\mathrm{i}\theta}}{2}\frac{b(r)}{r}\bar w = 0, \quad 0 < |z| < 1, \tag{16.4}$$

in which $w(z) = z^2 K^{1/4}(z)(\delta M + \mathrm{i}\delta L)$, and

$$b(r) = \frac{3\varphi'\varphi'' + \varphi\varphi''}{4\varphi''\sqrt{-\varphi\varphi''}}. \tag{16.5}$$

The sought–for function $w(z)$ is, just as before, expressed in terms of the Gaussian curvature K and the unknown variations δL and $\delta M (\delta N = -\delta L)$ of the coefficients of the second quadratic form L, M and N of the surface.

Let us dwell upon the properties of the real function $b(r)$. It is necessary to consider the right side of (16.5), using (16.2) as an implicit function of r. It is not hard to see that $b(r)$ is continuous for $0 < r \leq 1$. In order to characterize the behaviour of $b(r)$ at $r = 0$, we go over to the function $\xi = \omega(\varphi)$, using the expressions (16.1) and (16.3), in the right side of (16.5). Then in some neighbourhood of $r = 0$

$$b(r) = \beta \ln r + b_O(r),$$

where $b_O(r)$ is a continuous function. This formula is valid for the entire interval [0,1]. Moreover, it can be established that

$$\int_r^1 \frac{b_O(r)}{r}\,dr = a_O \beta \ln r + \frac{\beta+1}{2\beta} \ln|\ln r| + P(r), \quad r \to 0,$$

where $P(r)$ is a continuous function. Hence equation (16.4) takes the following form

$$\partial_{\bar{z}} w + \frac{\beta \ln r + b_O(r)}{2\bar{z}} \bar{w} = 0 \quad 0 < |z| < 1, \tag{16.6}$$

where $b_O(r)$ possesses the properties mentioned above.

In [3] the class of admissible infinitesimal bendings has been chosen in such a way that the variations $\delta L, \delta M, \delta N$ of the coefficients of the second quadratic form in conjugate isometric coordinates are continuously differentiable everywhere except at the pole ($z = 0$), where they are bounded. It has been established that in such class of deformations the surface is rigid, i.e. it does not admit non–trivial infinitesimal bendings.

Without dwelling upon the proof of this result, we should note: *infinitesimal bendings of the surface with an infinite order of flattening in the isolated flat point are described by the complex–valued equation (16.6), the coefficients of which have a polar singularity of order greater than 1.*

17 Unsolved problems

17.1 We use the new notation $O\xi\eta\zeta$ ($\xi = \rho\cos\psi, \eta = \rho\sin\psi$) instead of the previous rectangular coordinate system $Oxyz$. As was stated in §15, infinitesimal bendings of surfaces of type (15.2)

$$\zeta = \rho^n f(\psi) + R(\xi, \eta) \tag{17.1}$$

in the conjugate isometric parametrization are described by the unknown complex–valued function $w(z)$, which satisfies the equation (15.13):

$$\partial_{\bar{z}} w - \frac{\lambda(\varphi) + b_O(z)}{2\bar{z}} \bar{w} = 0 \quad (\varphi = \arg z). \tag{17.2}$$

Problem *what kind of supplementary demands should be imposed on the function $f(\psi)$ to get the relation*

$$\lambda(\varphi) \equiv \lambda_m e^{im\varphi}$$

where λ_m is a constant and m is an integer.

In this situation equation (17.2) takes the specific form which was analysed in ch.5.

17.2 If we exclude the particular cases $\lambda(\varphi) = \text{const}$ and $\lambda(\varphi) \equiv \lambda_m e^{im\varphi}$) then any results in relation to the solvability of (17.2) are unknown.

Apparently, the model object

$$\partial_{\bar{z}}\Phi - \frac{\lambda(\varphi)}{2\bar{z}}\bar{\Phi} = O, \tag{17.3}$$

although seeming to be simpler in comparison with (17.2), nevertheless remains fairly complicated for investigations.[3]

In this connection the following problem arises: *is it possible to construct a theory of equation (17.3) even if only in the following simple cases:*

1. $\lambda(\varphi) = \sum_{k=-m}^{m} \lambda_k e^{ik\varphi}$;
2. $\lambda(\varphi) = \lambda_O + \lambda_m e^{im\varphi}$;
3. $\lambda(\varphi) = \lambda_p e^{ip\varphi} + \lambda_q e^{iq\varphi}$.

Here $\lambda_O, \lambda_k, \lambda_m, \lambda_p, \lambda_q$ are complex constants and m, k, p, q are integers.

17.3 In chapter 5 we investigated the equation with coefficients having a polar singularity of order strictly greater than 1.

Problem *Is it possible to find a geometric realization of this case?*

---

[3] Recently certain progress in this respect has been made by A. Tungatarov; see 'On continuous solutions of the Carlemann–Vekua equation with a singular point', *Dokl. Akad. Nauk SSSR*, 1991, v. 319, no. 3.

Supplement
Generalized Cauchy–Riemann systems with a singular line

In this section we consider the equation

$$\partial_{\bar{z}}\Phi - \frac{e^{i\varphi}}{2}\frac{\lambda}{r-q}\bar{\Phi} = 0, \quad z \in G, \tag{1}$$

where $z = x + iy = re^{i\varphi}, i^2 = -1, \Phi(z)$ is an unknown complex–valued function, λ is a complex parameter, q is a positive number, G is a doubly–connected domain, one of the edges of G being the circle $L_O = \{z : |z| = q\}$ and the other is a simple closed contour L_1, enveloping L_O.

Equation (1) is a model equation in relation to

$$\partial_{\bar{z}}w - \frac{e^{i\varphi}}{2}\frac{b(z)}{r-q}\bar{w} = 0, \quad z \in G, \tag{2}$$

where $b(qe^{i\varphi}) = \lambda$. Equation (2) has arisen in investigating infinitesimal bendings of a surface of positive curvature with a conic point [38]. It proves to be of a more complicated nature in comparison to those equations which had been analysed in ch. 1–3 and which can be obtained from the present object with $q = O$. Nevertheless a unique method of investigation, comprising the singular equations of ch. 1–5, proves to be acceptable in this case as well.

In the present supplement our immediate aim is to construct elements of a theory of solutions to equation (1) similar to the model equations of ch. 2, 4 and 5. We restrict ourselves to considering the cases $|\lambda| < 1, 2|\lambda| \neq 1$ (further, we explain the reason of such restrictions).

It should be noted that the results, represented here, were obtained in [1].

1 The basic kernels of equation (1)

Let $E = \{z : |z| > q\}, z = x + iy = re^{i\varphi}$ and $\zeta = \xi + i\eta = \rho e^{i\theta}$. We examine a pair of functions in $E \times E$:

$$\Omega_1(z,\zeta) = \begin{cases} \Omega_1^*(z,\zeta), & |z| < |\zeta|, \\ -\Omega_1^*(\zeta,z), & |z| > |\zeta|, \end{cases} \tag{3}$$

$$\Omega_2(z,\zeta)=\begin{cases}\Omega_2^*(z,\zeta), & |z|<|\zeta|,\\ -\Omega_2^*(\zeta,z), & |z|>|\zeta|,\end{cases}$$

where

$$\Omega_1^*(z,\zeta)=\left(\frac{r-q}{\rho-q}\right)^{|\lambda|}\left[\frac{1}{2}+\sum_{k=1}^{\infty}\left(\frac{\rho}{q}\right)^{\beta_k}\left(\frac{r}{q}\right)^{k}\right.$$
$$\left.\times\frac{-I_k(\rho)F_{1k}(r)\mathrm{e}^{\mathrm{i}k(\varphi-\theta)}+J_k(r)F_{2k}(\rho)\mathrm{e}^{-\mathrm{i}k(\varphi-\theta)}}{2|\lambda|A_k}\right],$$

$$\Omega_2^*(z,\zeta)=\left(\frac{r-q}{\rho-q}\right)^{|\lambda|}\left[\frac{\lambda}{2|\lambda|}+\sum_{k=1}^{\infty}\left(\frac{\rho}{q}\right)^{\beta_k}\left(\frac{r}{q}\right)^{k}\right.$$
$$\left.\times\frac{|\lambda|^2F_{2k}(\rho)F_{1k}(r)\mathrm{e}^{\mathrm{i}k(\varphi-\theta)}-I_k(\rho)J_k(r)\mathrm{e}^{-\mathrm{i}k(\varphi-\theta)}}{2|\lambda|\bar{\lambda}A_k}\right].$$

Here

$$A_k=\frac{\Gamma(a_k+1-\beta_k)\Gamma(2|\lambda|)}{\Gamma(\gamma-\beta_k)\Gamma(a_k)},$$

$\Gamma(x)$ is the Eular gamma function, and

$$I_k(r)=(r-q)\frac{\mathrm{d}F_{2k}(r)}{\mathrm{d}r}+\left[\frac{r-q}{r}(\gamma-a_k-1)-|\lambda|\right]F_{2k}(r),$$

$$J_k(r)=(r-q)\frac{\mathrm{d}F_{1k}(r)}{\mathrm{d}r}+|\lambda|F_{1k}(r).$$

The hypergeometric functions are denoted by $F_{1k}(r)$ and $F_{2k}(r)$, [6] :

$$F_{1k}(r)=F(a_k,\beta_k,\gamma,-\frac{r-q}{q})$$

$$F_{2k}(r)=F(a_k+1-\gamma,1-\beta_k,a_k+1-\beta_k,\frac{q}{r}),$$

where

$$a_k=|\lambda|+k+\sqrt{k^2+|\lambda|^2},\qquad \beta_k=|\lambda|+k-\sqrt{k^2+|\lambda|^2},\qquad \gamma=2|\lambda|+1$$

and

$$F(a,b,c,z)=\frac{\Gamma(c)}{\Gamma(b)\Gamma(c-b)}\int_O^1\frac{t^{b-1}(1-t)^{c-b-1}}{(1-tz)^{\alpha}}\mathrm{d}t.$$

This representation for F_{2k} is valid only for $|\lambda|<1$.

The functions Ω_1 and Ω_2 are the basic kernels of equation (1). Their expressions were obtained according to the scheme of §1 ch.1, the restriction $2|\lambda|\neq 1$ being essential. We now state our assertion.

Lemma *The functions $\Omega_1(z,\zeta)$ and $\Omega_2(z,\zeta)$ are continuous in the two variables z and ζ in the direct product $E\times E$ except the set $z=\zeta$. For a fixed ζ (or z), not belonging to L_O and not equal to infinity, Ω_1 and Ω_2 have a zero of order $|\lambda|$ for $z=q\mathrm{e}^{\mathrm{i}\varphi}$ and $z=\infty$ (respectively for $\zeta=q\mathrm{e}^{\mathrm{i}\theta}$ and $\zeta=\infty$) and satisfy the relations*

$$\begin{cases} \partial_{\bar z}\Omega_1 - \dfrac{\lambda}{r-q}\dfrac{\mathrm{e}^{\mathrm{i}\varphi}}{2}\bar\Omega_2 = 0, \\ \partial_{\bar\zeta}\Omega_2 - \dfrac{\lambda}{r-q}\dfrac{\mathrm{e}^{\mathrm{i}\varphi}}{2}\bar\Omega_1 = 0, \end{cases} \qquad \begin{cases} \partial_{\bar\zeta}\Omega_1 + \dfrac{\bar\lambda}{\rho-q}\dfrac{\mathrm{e}^{\mathrm{i}\theta}}{2}\Omega_2 = 0, \\ \partial_{\bar\zeta}\Omega_2 + \dfrac{\lambda}{\rho-q}\dfrac{\mathrm{e}^{\mathrm{i}\theta}}{2}\Omega_1 = 0, \end{cases}$$

If G is a bounded domain then Ω_1 represented for $\zeta\in G$ in the form:

$$\Omega_1(z,\zeta) = \frac{\zeta}{\zeta-z} + \Omega_1^O(z,\zeta).$$

The following estimates for z and $\zeta\in G$ are valid

$$|\Omega_1^O| < C_1(|\lambda|,q,R,\varepsilon)\left\{\left|\frac{\ln|1-g_1(z,\zeta)|}{1-g_2(r,\rho)\mathrm{e}^{\mathrm{i}(\varphi-\theta)}}\right| + |1-g_1(z,\zeta)|^{-\varepsilon}\right\},$$

$$|\Omega_2| < C_2(|\lambda|,q,R)\left\{\left|\frac{\ln|1-g_1(z,\zeta)|}{1-g_1(r,\rho)\mathrm{e}^{\mathrm{i}(\varphi-\theta)}}\right|,\right.$$

where

$$g_1(z,\zeta) = \begin{cases} z/\zeta, & |z|<|\zeta|, \\ \zeta/z, & |z|>|\zeta|, \end{cases}$$

$$g_2(z,\zeta) = \begin{cases} q/\rho, & |z|<|\zeta|, \\ q/r & |z|>|\zeta|, \end{cases}$$

$R=\max|z|, z\in G, \varepsilon$ is an arbitrary positive fixed number, the constants C_1 and C_2 being bounded for values R near to q.

2 Cauchy generalized formula

Let $\bar G = G+L_O+L_1$. Let us consider solutions of equation (1) from the class $C(\bar G)\cap C_{\bar z}(\bar G - L_O)$, i.e. from the functions continuous in $\bar G$ and continuously differentiable outside L_O. On the basis of the Lemma, according to the scheme of §2 *ch.*2, we establish the equality

$$\frac{1}{2\pi\mathrm{i}}\int_{L_1}\frac{\Omega_1(z,\zeta)}{\zeta}\Phi(\zeta)\,\mathrm{d}\zeta - \frac{\Omega_2(z,\zeta)}{\bar\zeta}\overline{\Phi(\zeta)}\,\overline{\mathrm{d}\zeta}$$

$$= \begin{cases} \Phi(z) & \text{with } z\in G+L_O, \\ \frac{\alpha}{2}\Phi(z) & \text{with } z\in L_1, \\ 0 & \text{with } z\in E-G-L_1, \end{cases} \tag{4}$$

where $\alpha\pi$ is the interior angle of contour L_1 at the point z.

3 Analogy of Taylor series

Let $G = \{z : q < |z| < R\}$. Introducing $\zeta = Re^{i\gamma}$ into (4) for $z \in G + L_O$ and using the explicit forms of kernels Ω_1 and Ω_2, by means of term–by–term integration we get

$$\Phi(z) = \Big\{\big(a_O - \frac{\lambda}{|\lambda|}\bar{a}_O\big) + \sum_{k=1}^{\infty}\Big[\frac{-a_k + \lambda b_k}{|\lambda| A_k} F_{1k}(r)e^{ik\varphi} + \frac{\bar{\lambda} b_k - a_k}{\bar{\lambda}|\lambda| A_k} J_k(r)e^{-ik\varphi}\Big] r^k\Big\}(r-q)^{|\lambda|}, \tag{5}$$

where

$$a_O = \frac{(R-q)^{-|\lambda|}}{4\pi}\int_O^{2\pi} \Phi(Re^{i\gamma})\,d\gamma,$$

$$a_k = \frac{I_k(R)(R-q)^{-|\lambda|}R^{-k}}{4\pi}\left(\frac{R}{q}\right)^{\beta_k}\int_O^{2\pi} \Phi(Re^{i\gamma})e^{-ik\gamma}\,d\gamma,$$

$$b_k = \frac{F_{2k}(R)(R-q)^{-|\lambda|}R^{-k}}{4\pi}\left(\frac{R}{q}\right)^{\beta_k}\int_O^{2\pi} \Phi(Re^{i\gamma})e^{ik\gamma}\,d\gamma,$$

$$k = 1, 2, \dots .$$

Series (5) is the generalized Taylor series for solutions of equation (1) from the class $C(\bar{G}) \cap C_{\bar{z}}(\bar{G} - L_O)$.

The next statement issues from formula (5).

Theorem 1 *The following inequality*

$$|\Phi(z)| \le M_O(|\lambda|, q, R)(r-q)^{|\lambda|},$$

where M_O is a constant depending on $|\lambda|, q, R$, is valid for any solution of equation (1) from the class $C(\bar{G}) \cap C_{\bar{z}}(\bar{G} - L_O)$.

Theorem 2 *(Analogy of Liouville's theorem) If $\Phi(z)$ is a solution of equation (1) from the class $C(\bar{E}) \cap C_{\bar{z}}(\bar{E} - L_O)$ and is bounded in the closed domain $\bar{E}$ then* $\Phi(z) \equiv 0$.

Theorem 3 *A solution of equation (1) from the class $C(\bar{G}) \cap C_{\bar{z}}(G - L_O)$, possessing the property $\Phi(z) = o[|r-q|^{|\lambda|}], r \to q$, identically equal to zero.*

Theorem 4 *The inequality*

$$|\Phi(z)| < \frac{M_1(|\lambda|, q, R)}{R - r} \max_{z \in L_1} |\Phi(z)|,$$

where M_1 is a constant depending on $|\lambda|, q, R$ is satisfied for any solution of equation (1) from the class $C(\bar{G}) \cap C_{\bar{z}}(\bar{G} - L_O)$.

4 Analogy of Laurent series

Let a solution of equation (1) be examined in the circular ring $K = \{z : q < R_O < |z| < R_1\}$. Supposing $z \in K$ and integrating term–by–term in formula (4), we get

$$\begin{aligned}\Phi(z) =& \Big\{ (a_o - \frac{\lambda}{|\lambda|} \bar{a}_O) + \sum_{k=1}^{\infty} \Big[\frac{-a_k + \lambda \bar{b}_k}{|\lambda| A_k} F_{1k}(r) e^{ik\varphi} \\ &+ \frac{\bar{\lambda} b_k - \bar{a}_k}{\bar{\lambda} |\lambda| A_k} J_k(r) e^{-ik\varphi} \Big] \cdot r^k \Big\} (r - q)^{|\lambda|} \\ &+ \Big\{ (-c_O - \frac{\lambda}{|\lambda|} \bar{c}_O) + \sum_{k=1}^{\infty} \Big[\frac{-c_k + \lambda \bar{d}_k}{|\lambda| A_k} F_{2k}(r) e^{ik\varphi} \\ &+ \frac{\bar{\lambda} d_k - \bar{c}_k}{\bar{\lambda} |\lambda| A_k} I_k(r) e^{-ik\varphi} \Big] \cdot r^{-k} \left(\frac{r}{q} \right)^{\beta_k} \Big\} (r - q)^{-|\lambda|},\end{aligned} \tag{6}$$

where

$$c_O = \frac{(R_O - q)^{|\lambda|}}{4\pi} \int_O^{2\pi} \Phi(R_O e^{i\gamma})\, d\gamma,$$

$$c_k = \frac{J_k(R_O)(R_O - q)^{|\lambda|} R_O^k}{4\pi} \int_O^{2\pi} \Phi(R_O e^{i\gamma}) e^{-ik\gamma}\, d\gamma,$$

$$d_k = \frac{F_{1k}(R_O)(R_O - q)^{|\lambda|} R_O^k}{4\pi} \int_O^{2\pi} \Phi(R_O e^{i\gamma}) e^{-ik\gamma}\, d\gamma,$$

$$k = 1, 2, \ldots$$

a_O, a_k and b_k are the same quantities as in formula (5).

The right side of formula (6) consists of two series: the first one contains the multiplier $(r - q)^{|\lambda|}$, and the second one contains $(r - q)^{-|\lambda|}$. If $\Phi(z)$ is absolutely integrable on the contours L_O and L_1, then both series simultaneously converge absolutely and uniformly in the ring $R_O < |z| < R$.

5 Generalized integral of Cauchy type

Let us consider the function $\Phi(z)$, defined by the integral

$$\Phi(z) = \frac{1}{2\pi i} \int_{L_1} \frac{\Omega_1(z,\zeta)}{\zeta} \nu(\zeta)\, d\zeta - \frac{\Omega_2(z,\zeta)}{\bar{z}} \overline{\nu(\zeta) d\zeta}, \tag{7}$$

where Ω_1 and Ω_2 are given by equality (3) and $\nu(\zeta)$ is a function, Holder continuous on L_1. This integral determines the functions $\Phi^+(z)$ and $\Phi^-(z)$ for $z \in G$ and $z \in E - G - L_1$ respectively. They are solutions of equation (1). In addition $\Phi^+(z) \in C(G + L_O + L_1) \cap C_{\bar{z}}(G)$ and $\Phi^-(z) \in C_{\bar{z}}(E - G - L_1), \Phi^-(\infty) = 0$.

Integral (7) exists in the sense of the principal value for $z = z_O \in L_1$ and defines the function $\Phi(z_O)$ satisfying the Holder condition.

For $z \to z_O, z \in G$ (or $z \in E - G - L_1$), we get an analogy of the Sohockii–Plemelj formulas:

$$\Phi^+(z_O) = \frac{1}{2}\nu(z_O) + \Phi(z_O), \qquad \Phi^-(z_O) = -\frac{1}{2}\nu(z_O) + \Phi(z_O).$$

References

1. ABDUŠUKUROV A.M., USMANOV Z.D.

   On the equation $2(r-q)\partial_{\bar z}\Phi - \lambda e^{i\varphi}\bar\Phi = 0$. *Izv. Akad. Nauk Tajik SSR* 1977, no.2(64), 66–70 (Russian). MR 58#11 449.

2. ACIL'DIEV A.I., USMANOV Z.D.

   Rigidity of a surface with a point of flattening. *Math. USSR sbornik*, 1967, v.2, no.1, 77–83.

3. AHMEDOV R.

   Investigation of solutions of an elliptic system in a neighbourhood of a singular point. *Dokl. Akad. Nauk Tajik. SSR*, 1986, v.29, no.2, 67–70 (Russian).

4. AHMEDOV, R.

   Boundary value problems for the equation $2\bar z\partial_{\bar z}w - (\lambda e^{in\varphi} + b(z))\bar w = F$. *Izv. Akad. Nauk Tajik SSR*, 1987, no.1, p.32, dep. VINITI no. 208-I 87 (Russian).

5. BARKHIN G.S., FOMENKO V.T.

   On bendings of positive curvature surfaces with a boundary. *Sibirsk. Math. Z.* 1963, v.4, no.1, 32–47 (Russian).

6. BATEMAN H., ERDELYI A.

   Higher transcendental functions, II. McGraw Hill, New York, Toronto, London 1953.

7. BERNSTEIN S.N.

   Study and integration of 2nd order partial differrential equations of elliptic type. *Sobranie sochineni*, 1960, v.3, 24–110 (Russian).

8. BERRY R.J.

On an integral invariant of 4th degree binary forms. *Uspehi Mat. Nauk*, 1952, v.7, 3(49), 125–130 (Russian).

9. BESOV O.V., IL'IN V.P., NICOL'SKY S.M.

Integral reprsentations for functions and inclusion theorems, Moscow Nauka, 1975, p.480 (Russian).

10. BLIEV N.K.

On the existence of an analytic solution for an elliptic system degenerating in a point *Izv. Akad. Nauk Kasah. SSR*, series fiz.-mat. nauk, 1965, no.1, 96–104 (Russian).

11. BLIEV N.K.

On the existence of analytic solutions for a singular elliptic system *Vestnik Akad. Nauk Kasah. SSR*, 1965, no.4, 74–79.

12. BLIEV N.K.

On necessary and sufficient conditions for the existence of analytic solutions of a singular system *Izv. Akad. Nauk Kasah. SSR*, series fiz.-mat. nauk, 1967, no.1, 93–95 (Russian).

13. BLIEV N.K.

Generalized analytic functions in fractional spaces. *Alma-Ata, Nauka*, 1985, p.159 (Russian).

14. BLIEV N.K., TUNGATAROV, A.

On a generalized Cauchy–Riemann system with a singular point. Differential equations and their application, Alma-Ata, Nauka, 1975, 29–36.

15. DORFMAN A.G.

Solvability of bending equations for some classes of surfaces. *Uspehi Mat. Nauk*, 1957, 12:2(74), 147–150 (Russian).

16. EFIMOV N.V.

Bending of a neighbourhood of a surface parabolic point. *Mat. sbornik*, 1939, v.6(48), 427–474 (Russian).

17. EFIMOV N.V.

Proof of the existence of a rigid surface in the small. *Dokl. Akad. Nauk SSSR*, 1940, no. 27, 314–317 (Russian).

18. EFIMOV N.V.

Study of bendings of a surface with a flat point. *Mat. sbornik*, 1946, v.19(61): 3, 416–488 (Russian).

19. EFIMOV N.V.

Rigidity in the small *Dokl. Akad. Nauk SSSR*, 1948, v.60, no.5, 761–764 (Russian).

20. EFIMOV N.V.

Qualitative questions of a theory of surface deformations. *Uspehi Mat. Nauk*, 1948, v.3, 2(24), 47–158 (Russian).

21. EFIMOV N.V.

Qualitative questions of a theory of surface deformations in the small. *Trudi Mat. ins. Akad. Nauk SSSR*, 1949, v.30 1–128, (Russian).

22. EFIMOV N.V., USMANOV Z.D.

Infinitesimal bendings of a surface with a point of flatness. *Soviet Math. Dokl.*, 1973, v.14, no.1. 22–25.

23. GAHOV F.D.

Boundary value problems. Moskow, Fizmatgiz, 1963, p.640 (Russian).

24. GRADSTEIN I.S., RIGIK I.M.

Tables of integrals, sums, series and products. Moscow, Fizmatgiz, 1963, p.1100 (Russian).

25. HOESLI R.I.

Spezielle Flachen mit Flachpunkten, Compositio Mathematica, 8 fasc.2 (1950).

26. INCE E.

Specific functions. Moscow, Nauka, 1968, p.344 (Russian).

27. KAGAN V.F.

Basis of a surface theory, part 1. Moscow, Gostehizd, 1947, p.511 (Russian).

28. KAGAN V.F.

Basis of a surface theory, part 2. Moscow, Gostehizd, 1948, p.407 (Russian).

29. KANTOROVICH L.V., AKILOV G.P.

Functional analysis in normed spaces. Moscow, Fizmatgiz, 1959, p.684 (Russian).

30. LAVRENT'EV M.A., SCHABAT B.V.

Methods of a theory of complex variable functions. Moscow, Fizmatgiz, 1958, p.678 (Russian).

31. MAKAROVA Z.G.

Study of the N integral invariant of $n > 4$ degree binary forms. *Mat. sbornik*, 1953, v.33(75), 233–240 (Russian).

32. MAKATSARIA G.T.

The Carleman–Vekua equation with a polar singularity in the coefficients Sbornik Boundary value problems of generalized analytic functions and their applications, Tbilisi State University, Institute of applied mathematics, 1983, 61–78 (Russian).

33. MAKATSARIA G.T.

Theorems of Liouville type for generalized analytic functions. *Soobschenia Akad. Nauk Georg. SSR*, 1984, v.113, no.3, 485–488 (Russian).

34. MARKUSCHEVICH A.I.

A course of a theory of analytic functions. Moscow, Gostehizd, 1957, p.335 (Russian).

35. MIKHAILOV L.G.

Integral equations with a-1 degree homogeneous kernel. Izd Donish, Dushanbe 1966, p.48 (Russian).

36. MIKHAILOV L.G.

A new class of singular integral equations and its application to differential equations with singular coefficients. Hoord Hoff Publishing House, Netherlands, Groningen, 1970.

37 MIKHAILOV L.G.

On some integral equations of the theory of generalized analytic functions in a singular case. *Soviet Math. Dokl.* 1970, v.11, no.2., p.531–534 (Russian).

38. MIKHAILOV L.G., USMANOV Z.D.

Infinitesimal bending of a surface of revolution with positive curvature and with a conical or parabolic point at the pole. *Doklady*, tom 166, no. 4, 1966, 174–177.

39. MUSHELISVILI N.I.

Singular integral equations. Noord hoff, Groningen, 1953.

40. NAZIROV G.

Analytic solutions of some elliptic equations with singular coefficients. *Dokl. Akad. Nauk Tajik SSR*, 1961, v.4, no.3, 3–6.

41. NAZMIDDINOV H.

Boundary value problems for a model generalized Cauchy–Riemann system with more than 1st order point singularity in the coefficients. *Izv. Akad. Nauk Tajik SSR*, 1975, no. 4 (58), 72–75 (Russian).

42. NAZMIDDINOV H.

Study of the equations $2|z|^{1+2\nu}\partial_{\bar{z}}\Phi - \lambda e^{i(n+1)\varphi}\Phi = 0$. *Izv. Akad. Nauk Tajik SSR*, 1976, no.2(60), 100–104 (Russian).

43. NAZMIDDINOV H., USMANOV Z.D.

A generalized Cauchy–Riemann system with more than 1st order point singularity in the coefficients *Dokl. Akad. Nauk Tajik SSR*, 1975, v.18, no.5, 11–15 (Russian).

44. POGORELOV A.V.

Infinitesimal bendings of convex surfaces. Kharkov, Izd. of Kharkov University, 1958, p.147 (Russian).

45. SABITOV I.H.

Study of rigidity of analytic rotation surfaces with a flat point at the pole. *Vestnik MGU*, series 1, 1986, no.5, 29–36 (Russian).

46. SMIRNOV V.I.

A course of higher mathematics, v.3, part 2. Moscow, Fizmatgiz, 1969, p.672 (Russian).

47. SOBOLEV C.L.

Equations of mathematical physics. Moscow, Nauka, 1968, p.443 (Russian).

48. STOILOV S.

A theory of functions of a complex variable, part 1. Moscow, Inostran. lit., 1962, p.364 (Russian).

49. TARTAKOVSKY V.A.

On the N–invariant of Efimov from a theory of surface bendings. *Mat. sbornik*, 1963, 32 (74), 225–248 (Russian).

50. TUNGATAROV A.

On some class of irregular generalized Cauchy–Riemann systems *Izv. Akad. Nauk Kasah. SSR*, series fiz.-mat. nauk, 1977, no.5, 56–61 (Russian).

51. TUNGATAROV A.

Some irregular cases of generalized Cauchy–Riemann systems and a study of the Riemann–Hilbert boundary value problem. Candidate dissertation, Alma-Ata, 1978, p.97 (Russian).

52. USMANOV Z.D.

An investigation of the model generalized Cauchy–Riemann systems with singular coefficients. *Dokl. Akad. Nauk Tajik. SSR* 1968, v.11, no.11, 6–10 (Russian). MR 41 # 2034.

53. USMANOV Z.D.

On infinitesimal deformations of surfaces of positive curvature with an isolated flat point *Math. USSR sbornik* 1970, v.12, no.4, 595–614, MR 42 # 8428.

54. USMANOV Z.D.

An integral equation for a certain class of generalized analytic functions with a fixed singular point. *Dokl. Akad. Nauk Tajik. SSR* 1971, v.14, no.10, 12–15 (Russian). MR 45# 3730. 30A92.

55. USMANOV Z.D.

The Dirichlet problem for a generalized Cauchy–Riemann system in the singular case. *Dokl. Akad. Nauk Tajik SSR* 1971, v.14, no.11, 16–20 (Russian). MR 45 # 3731.

56. USMANOV Z.D.

The Riemann–Hilbert problem for a certain class of generalized analytic functions with a fixed singular point, I. *Izv. Akad. Nauk Tajik SSR*, 1971, v.41, no.3, 10–14 (Russian). MR 45 # 3732. 30A92.

57. USMANOV Z.D.

On the theory of the equation $2\bar{z}\partial_{\bar{z}}\Phi - \lambda\bar{\Phi} = 0$. *Dokl. Akad. Nauk Tajik SSR* 1972, v.15, no.5, 13–16. (Russian) MR 46#9364. 130A88.

58. USMANOV Z.D.

An effective solution of a certain two–dimensional integral equation with homogeneous kernel and its applications. *Differencial'nye uravnenija*, 1972, v.8, no.12, 2267–2270, (Russian). MR 47# 7357. 45E05 (30A88).

59. USMANOV Z.D.

The Riemann–Hilbert problem for a certain class of generalized analytic functions with a fixed singular point, II. *Dokl. Akad. Nauk Tajik SSR*, 1972, v.15, no.4, 10–13. (Russian). MR 46 # 9363. 30A88.

60. USMANOV Z.D.

Junctions of generalized analytic functions with a fixed singular point. *Dokl. Akad. Nauk Tajik. SSR*, 1972, v.15, no. 16, 13–17 (Russian). MR 48 # 6442. 30A88.

61. USMANOV Z.D.

Studies of generalized analytic functions with a fixed singular point. *Izv. Akad. Nauk Tajik SSR*, 1972, v.43, no.1, 13–6 (Russian). MR 47 # 5268.

62. USMANOV Z.D.

On a problem concerning the deformation of a surface with a flat point. *Math. USSR sbornik*, 1972, v.18, no.1, 61–81. MR 47 #2532.53E45.

63. USMANOV Z.D.

A certain class of generalized Cauchy–Riemann systems with a singular point. *Sibirsk. Mat. Z.*, 1973, v.14, no.5, 1076–1087. (Russian). MR 49#5371. 30A92.

64. USMANOV Z.D.

An investigation of the equations of the theory of infinitesimal bendings of surfaces of positive curvature with the point of flattening. Tbilisi University, Seminar of institute of applied mathematics, Abstracts of papers 1973, no.8, 85–87 (Russian). MR 51 # 11379. 53E45.

65. USMANOV Z.D.

The structure of the zeros of generalized analytic functions with a fixed singular point. *Izv. Akad. Nauk Tajik.* 1974, no.3 (53), 7–10 (Russian). MR 54# 2988.30A92.

66. USMANOV Z.D.

On the question of rigidity in the small in the class C^∞ *Mat. Zametki*, 1977, v.22, no.5, 643–648 (Russian), MR 58 # 12862.

67. USMANOV Z.D.

Infinitesimal bendings of surfaces of positive curvature with a flat point. Diff. geometry, Banach Centre Publications, Warsaw, 1984, no.12, 241–272 (Russian).

68. USMANOV Z.D.

Conjugate isomeetric parametrization of a surface in a neighbourhood of a point of flattening. Seminar of Institute of Applied Mathematics, Tbilisi, 1985, v.1, no.1, 205–208 (Russian).

69. VEKUA I.N.

Generalized analytic functions. Pergamon Press, London, Addison–Wesley, Reading, MA, 1962, MR 21 # 7288; 27 # 321.

70. VEKUA I.N.

Stationary singular points of generalized analytic functions. *Soviet Mat. Dokl.*, 1962, 3, 917. MR # 217.

71. VEKUA I.N.

On the class of elliptic systems with singularities. Proc. of the International Conference on Functional Analysis and Related Topics, Tokyo, April, 1969.

72. VILENKIN N.J.

Specific functions and a theory of group representations. Moscow, Nauka, 1965, p. 588. (Russian).

73. VOYTSEHOVSKY M.I.

Rigidity of convex surfaces with a boundary. *Vestnik MGU*, 1964, series 1, no.6, 35–40 (Russian).

74. WATSON G.N.

Theory of Bessel Functions, part 1. Moscow, Internat. literature, 1949.